Practical Electronics and Arduino in 8 Hours

2020 edition

Jim Fragos

Copyright © 2020 by Jim Fragos (Dimitrios Fragkos)
All rights reserved. This book or any portion thereof
may not be reproduced or used in any manner whatsoever
without the express written permission of the publisher
except for the use of brief quotations in a book review.

Printed in the United States of America

First Printing: December 2020

Printed book ISBN: 9798580074351
Electronic book: ASIN number on Amazon only

Dimitrios Fragkos (Jim Fragos)
For permissions to re-print information of this book or related
to that business, you are very welcome to reach the author by
email at: jfragos@jfragos.com
web site: www.jfragos.com

To my beloved wife Sophie,

For a million good reasons, one of which is her support to write this book

Contents

1. Fly over the electronics planet 1

1.1 The elements: Voltage and Current 1

1.2 Resistance: the 80% of all of the ingredients 7

1.3 Switches, LEDs and some connection topologies to play with 10

1.4 Passive components fly over 18

1.5 Active components and ICs fly over 24

1.6 Components connecting techniques 29

1.7 Regulators – a first glimpse to ICs 33

1.8 One plus one makes 10: the binary system and the digital world 39

1.9 Microcontroller anatomy: Exploration of the magic castle 43

1.10 Arduinos and the Arduino UNO board 49

1.11 Sensors and things that move stuff and display stuff 57

1.12 Programming: the big picture and one easy program 73

1.13 We just circumnavigated the electronics planet, lets land and do that again on the ground with a fast car 82

2. Drive fast through the electronics Wonderland 83

2.1 Voltage and Current real engineering 83

2.2 Resistors recipes 89

2.3 Components: Technologies, sizes and where to find them 93

2.4 PCBs, soldering techniques and equipment 106

2.5 Further "must-have" knowledge on passive components and signals . 121

2.6 Active components and ICs: Regulators and other useful 133

2.7 Analog signals and measuring instruments 143

2.8 Microcontroller anatomy: Deeper exploration of the rooms of the magic castle 153

2.9 The hardware of the Arduino UNO board and others even greater .. 161

2.10 Some real Arduino circuits with sensors and displays 168

2.11 All right, it's quite embarrassing to ask: What is REALLY a computer? 176

2.12 C++ introduction for the non-programmer 180

2.13	More Arduino programming	187
2.14	A few simple programs to play with	196
2.15	Wireless communications and Internet engineering	212
2.16	Fun has just begun, where to go next	218

3. Appendix 224

3.1	Multiplier prefixes	224
3.2	A real datasheet of an LED	225
3.3	Datasheet highlights of some notable MCUs	229

Whom this book is for

Anyone who is in high school up to a Ph.D. owner in any field. The level of the mathematics and the physics needed for background is almost zero. Engineering taught in here is kept at a pictorial level, math is avoided when not needed. People who are at their first steps in electronics already, will also benefit a lot from this book.

What can you expect to learn?

- Meet electronics. They will flirt you and maybe you will fall in love
- **Speak the language** of Electronics & Embedded Systems Engineers
- **Understand** the most needed concepts of hardware and software in deep level, from the ground - up
- Gain **applied** knowledge for real-world electronic components of the **latest technology**
- Practical **assembly techniques, measuring techniques** and **lab equipment** are covered
- Understand what a **microcontroller** is and get your hands on the one inside the Arduino Uno board
- Make your simple **programs** and understand simple programs made by others
- Understand most of the electronics connection diagrams (**schematics**) of Arduino projects. You also may detect imperfections in some of them!
- **Make** electronic circuits of your design with self-guided further reading

All **understanding** will be at a level, amazingly, not of a beginner, but of an intermediate+ embedded systems hobbyist. **The purpose of this book is to start you up.** Your foundations built here will be placed on the latest of the electronics technology (as of 2020). Next is diving into the knowledge yourself in the process of making your projects. You will be able to do that. While building experience, your first homebrew projects may be trial and smoke until they work. But this is the self-teaching way, the best to my opinion. You will need lots of hundreds of hours to practice knowledge, to gain experience and to soak into details in order to make a real product for the market. This book will try to give you a ticket to that. The traveling is up to you.

Learning approach

This book will approach teaching as an adventurous, exploring journey in two phases for letting more of this knowledge ender into your long term memory. Phase one: Get the big picture first. Phase two: Revisit all and go into more details. Equations and mathematical theory are avoided like hell throughout all of the book, they are little needed anyway for the 95% of the simple electronic projects. Pictorial intuition for understanding the cause-and-effect and how to avoid most common mistakes is where the focus is on.

About the author

The author owns a Physics degree and a Ph.D. in Virtual Reality technology, but those are overtaken by his love for electronics that started at the age of 11 as a self-taught hobbyist. In the last 18 years he is a professional embedded systems designer. He has made on his own more than 10 commercial or industrial products and has designed more than 40 worthy electronics projects where design, prototyping, programming and testing was performed solely by him. During his carrier within companies, he has trained many junior engineers. That spawned the experience and the material to write this book. At the time of writing he is the chief engineer for the design of educational robots in an awesome company.

You can ask your questions while reading, in a forum created for the book. "Google" the title of the book + the word "forum" and you will get there easily enough. I will be happy to read words from you while you read mine.

Welcome aboard

Let's get to it! Imagine the knowledge of embedded electronics is a (round) planet where all civilization is at the equator. We will take off with a supersonic airplane from some point, see it all flying fast from above and get the first overall picture. We will land at the same airport at the same direction we took-off and we will then get on a super-fast car. We will drive fast through almost all the scenery, looking it from a lot closer, but yet, we will always move fast, not having time to see anything in its full analysis. We will see the fundamental concepts and the big ideas behind each matter in order to enable our intuition to play with it, not words and terminology only. Hope lots of fireworks will explode inside your brain.

We will not follow the latest trend "we will make project X" since the project-driven approach is not going to cover the majority of the concepts needed to understand and conceive hundreds of projects out there waiting for you. You should show a little patience to read a classic kind of book with the motive: I know $\neq$ I understand and highly unorthodox in regards to what the "must-have" electronics knowledge is. We will go from 0 to X with no gaps, you are welcome to skip parts you already know.

In order to better digest it (with your memory, not the stomach) it is highly advised to break the reading of this book to four sessions at least. The "8 hours" on the title suggests an average total reading time with very small pauses to research or think in between.

<u>A (boring) Legal Notice (that is unlikely to happen):</u>

This book serves as a teaching aid only. Any particular application resulting in damage of property, loss of property or health damage based on the information given in this book will not hold the author or the publisher liable. We make no warranties, express or implied, that the examples, data, or other information in this book are free of error.

In simple words, you should be careful at all (rare) cases when working with dangerous voltages or components. In such situations you should know how to protect yourself. If anything breaks, you keep both pieces.

Electronics is not a dangerous hobby in general. It is relatively not expensive either. It involves brain and art in and out of a computer screen, it is, in all of its phases, exciting for all who get to fall in love with it.

Fasten your seatbelts, takeoff will have some "G"s of acceleration.

1. FLY OVER THE ELECTRONICS PLANET

1.1 THE ELEMENTS: VOLTAGE AND CURRENT

WIRES

Electronics are about electrons! That wise phrase said, the next question should be: what are electrons? We physicists, can tell you with much pride that: we do not have a damn clue what they are! Really zip, not to the least. But, we know amazingly precisely most - if not all - of their properties like their mass, charge etc. Let's approach their description and their function in the context of practicality for building circuits only.

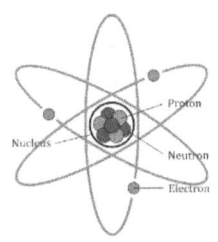

All materials are comprised of 3 particles. Protons, neutrons, and electrons. Together they form atoms. The number of protons (which are positively charged) equals the number of electrons (negatively charged), so they are electrically neutral. That number defines what element the material is (there are 118 known today some of them not stable). Some of the elements are called metals. Metals are known for conducting electricity.... What is that?

Electrons are **charged**. They carry negative electric charge that attracts (or is attracted by) positive charges and is repelled by negative charges.

In a rigid metal body (or a wire) some of their atom's electrons move almost freely around all the metal's body. **Electricity** is about electrical charges (electrons) motion or accumulation of excess negative charges in an area (more negative electrons than positive protons) or the opposite.

Note that all this pictorial description made here is oversimplified but it does our job for the rest of this book. The truth has much quantum physics involved and we need some books to describe it fully. Let's quickly see what the **electric current** (oversimplified) is:

CURRENT

Here we will make a picture which may be way too oversimplified but will explain and visualize in our mind what the current is: We will only talk about metals, the electricity conductors. In metals as said, some of their atoms' electrons move freely. Let's imagine a

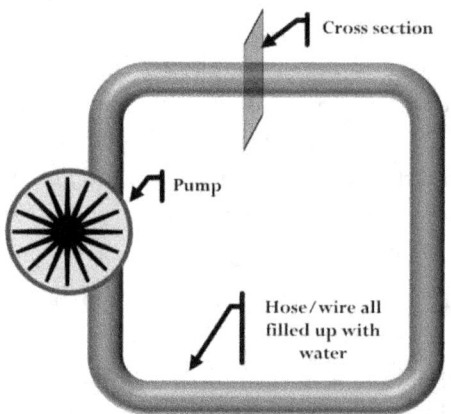

metal wire like a hose inside which, instead of water, electrons flow, but like water they make an uncompressed fluid. Also imagine that all of the hose/wire is filled up totally by this uncompressed fluid with no gaps inside it. This description is called "The electronic–hydraulic analogy". Our total water-like fluid volume (that is the total volume of the metal) has no importance. Important is the flow rate that is how much fluid passes totally by a cross section per time. No matter how wide or narrow that part of the "hose" is, we care about how many liters per second pass through there. Back to electrons reality, the volume passed is the number of electrons that is an amount of electric charge. Electric charge Q is measured in Coulomb units (one Coulomb is 6241509129000000000 or $6.241509129 \cdot 10^{18}$ electrons), so, electric current is:

$$I_{(Amperes)} = \frac{Q_{(Coulombs)}}{t_{(sec)}}$$ Where Q is the total charge passed through our section in Coulomb units, t is the time period in seconds we have been counting charge passing through (double time, double Q, same result) and I is the current. Current is measured in Amperes (A) due to the inventor's name, one Ampere is flaw rate of one Coulomb per second. Do not worry about mathematics for the rest of the book. They will be kept at analogies only, like the previous equation, easy to remember and as an engineer, you will be inclined to derive rather than remembering them.

One Ampere of current is medium. It requires about 0.5mm wide wire made of copper to pass through without blowing or heating the wire. In low voltages (we will come next to that), if shorted, it will make a negligible but just noticeable spark. But for the most of our circuits, since they are low power, it is considered a rather big amount of current, usually an Arduino requires less than 0.1A total supply current without any motors connected to it and a small motor needs about half to two Amperes (Amps). A USB socket of a computer provides 0.5A maximum current. Our car battery provides about 10A to light the front headlights and that takes quite thick wires.

What the current can do:

- ➤ It heats up things (the most common failures are due to materials melting by excessive current that flows through them)
- ➤ It can produce light
- ➤ It produces magnetic field around the wire it flows through. Motors and loudspeakers work using that magnetic repulsion / attraction.
- ➤ It is the "fuel" needed for circuits to work, usually termed as current consumption.

We may say that if a circuit has veins, current is the blood…

Last… Current has direction of flow, usually displayed with an arrow

VOLTAGE

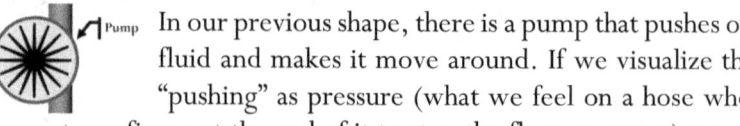 In our previous shape, there is a pump that pushes our fluid and makes it move around. If we visualize that "pushing" as pressure (what we feel on a hose when we put our finger at the end of it to stop the flow on water) we get the concept of the Voltage. Voltage is roughly how much "tension" there is to move electrons. It is measured in units of Volts (Alessandro Volta) and is present as a "tension" or "pressure" even if there is no wire (hose) connected (to the pump). We may think that the pump never stops rotating, never turns off.

One Volt (V) is a small "voltage" but very easy to measure precisely enough (e.g. to know if it is within 0.999V to 1.001V) with very common instruments. Most of the circuits using microcontrollers work with 3V to 5V voltage. We may start feeling the slightest of an electric shock if we touch over 30V with our fingers.

Current as we said earlier, is the flow rate. Voltage is the "pressure difference" between **two points** and may be produced by:

- ➤ A battery that uses chemical reactions
- ➤ An electrical generator that alternates magnetic field intensity around wires
- ➤ Stored electrical field (in our body in dry atmosphere when we rub with woolen clothes, or in charged capacitors)
- ➤ From light using photovoltaic cells

And other means.

Voltage has no direction (as current has) but one of the two of its points (poles) has higher "pressure" and the other has "lower" and they are symbolized with the + sign for the higher and the − sign for the lower. We may think it as "sucking" fluid from the "-" point and expelling it from the "+" using the pump analogy.

Circuits

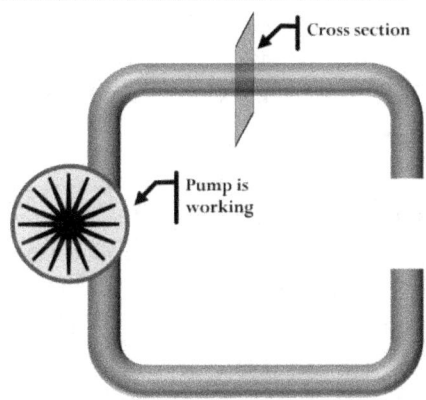

In our electronic–hydraulic analogy, our fluid needs a closed loop hose to flow since the pump in our previous shape only recycles the fluid inside our hose. A wire that makes a closed loop is called circuit or **closed circuit**. **Open circuits** have zero current flowing through them. Voltage is what causes current (the flow). The more it is, the higher the current is (the flow rate) if flow can be made at all (if e.g. there is a closed wire (hose) around the two points of Voltage). Voltage usually does nothing if there is nothing connected to it to make a closed circuit.

Let's stop drawing pipes and move on to how we draw circuits.

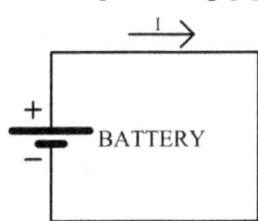

Wires are drawn by simple lines and each of the components connected (like the pump previously) have their own symbol. For instance the symbol of a real Voltage generator (i.e. a battery) within the same circuit as the one discussed earlier is the one on the left. We draw circuits that way for understanding to the best what is connected to what. These are called schematic diagrams or just schematics.

A few very important things about Current and Voltage

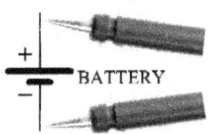

In order to measure voltage we just touch the two probes of a Voltmeter at the two poles of our Voltage source. We will get to measuring instruments and techniques on chapter 2.7.

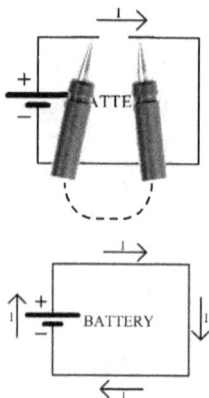

In order to measure current we have to break our circuit and by-pass the flow of electrons through our instrument (Current meter) which acts like a wire

In the circuit on the left the current is the same at any point we measure it. That's because the "electrons fluid" is uncompressed so it cannot be accumulated at any area of our circuit neither it can "evaporate" or be lost in areas. The same number of electrons pass every second from any cross section of it.

The current flowing in a circuit may be split and re-combine. Its

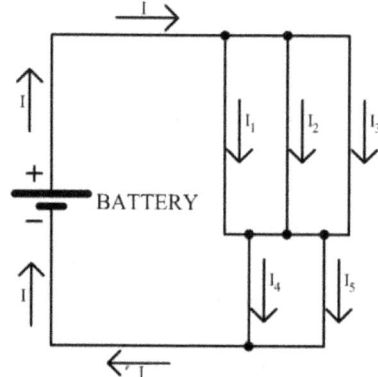

non-compressible and lossless nature makes it like an ideal river of water which may split to branches and recombine. When recombined, all the water of the (ideal) river should have exactly the same flow rate as before the branching. In the circuit on the left, after flowing through the branches, the current becomes the same as before. At the branches it splits in a way that no flow is lost, that is $I_1+I_2+I_3 = I$ and $I_4+I_5 = I$

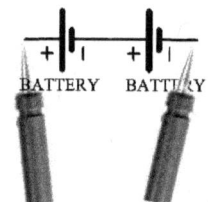

Voltages add-up when connected in series, the "+" to the "-", so they push electrons to the same direction. Two batteries 1.5V each in series make a voltage source of 3V. If voltage sources are connected opposite sides ("+" to "+") they cancel each other and the total voltage is the subtraction of their voltages.

So far we said about almost nothing "touchable" but rather "spooky" things about invisible electrons. Hung on, components are soon coming.

1.2 Resistance: the 80% of all of the ingredients

In our good electronic–hydraulic analogy, the pump (voltage) is what pushes the fluid. Imagine now that the hose or hoses it passes through are not very wide, so they provide (according to how thin they are) resistance in the flow. The flow rate then should be higher, the higher the pressure (voltage) is and lower the higher the total resistance is. A German guy called Georg Ohm said in 1826 that current is proportional to the voltage (e.g. doubling the voltage doubles the current) and reversely proportional to the resistance (e.g. doubling the resistance halves the current). This is the ultimately most important law in electronics and is expressed as:

$$\text{Ohm's law:} \quad I = \frac{V}{R}$$

We will fall on this so simple law (equation) in almost every page of the rest of this book, resistance is everywhere in electronics. Ohm's law correlates 3 things, current, voltage and resistance, so if we know the resistance we know the relation between current and voltage and if we know any two of those we derive the third. We may re-state Ohm's law as $V = IR$.

The measurement unit of resistance is the Ohm (or Ω that is the Greek Omega). We may make the last re-arrangement of the Ohm law in the form $R = \frac{V}{I}$ and say that 1Ohm equals to that resistance that makes 1A of current flowing through it when 1V of voltage is applied across it.

Resistors

Resistors are the most common component in electronics. Their purpose is to provide exactly the amount of resistance we deliberately need at any point in a circuit. They come in their older form (from around 40's to 80's called **"through hole"**)

of a component with two wires and in their more recent form (about 2-3 last decades, called **SMD**) of surface mounted components. We can find resistors from less than 0.1 Ohm to 10,000,000Ohms values very easily. They also are of the stuff to do the happiest shopping for, since they are the cheapest of all electronics components, so ridiculously cheap that each costs less than a cent usually. Intentional usage of resistance (resistors placement) is often used to: limit current, set current to a value we want, reduce voltage and many-many other functions we will be seeing throughout the book.

Kilo, Mega, Giga

An average resistor's value is about 10,000 Ohms. Big ones are 1,000,000 Ohms and over. In order to avoid writing and counting all those zeros you make take a look at the appendix 3.1 for the Kilo, Mega and Giga prefixes. For example 2400 Ohms is written handily as 2.4K Ohms or 2.4K or 2K4 for avoiding mistakes of not noticing the dot.

Unintentional resistance

There are times where we meet resistance where we do not want to. Unfortunately resistance is practically everywhere, whether we want it or not depends on how much it is. Every metal, or wire has some unwanted resistance, very small, and it is higher the longer it is, lower the wider it is. 1mm wide 1 meter long copper wire for example has 0.021 Ohms resistance. Resistance is also found in almost all electronic components as an unwanted parameter (sometime in the region of 10s of Ohms). Many components or sub-systems behave as resistors, like lamps, loudspeakers, heaters and others, where their resistor is considered as "load". The less Ohms it is the more current is required to flow through that component to work well or the more current it "consumes" for providing its functionality.

CONNECTING RESISTORS TOGETHER

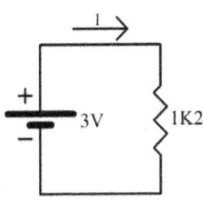

On the left we see a schematic comprising of one resistor only. The schematic symbol of the resistor this: —⟋⟍⟋⟍— but rarely used is also this one: ─▭─ In the particular circuit we can calculate the current I (the same passing through all of this loop of wires, battery and resistor) by the Ohm's law, as I=V/R = 3V/1200Ohm = 0.0025A (or 2.5mA per Appendix 3.1).

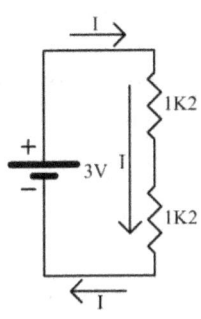

In the circuit on the left, we have two resistors in series. Resistors in series act like one resistor. How big should that be? Well, the current flow meets one narrow path and then another narrow path, so the total resistance is 1.2K+1.2K = 2.4K. **Resistors in series add up.** Applying Ohm's low using that total resistance we find the current I to be I=3V/2400Ohm = 0.0015A (or 1.5mA), yes, the half.

Let's see what resistors connected in parallel do. Current I will branch into I_1 and I_2 ($I=I_1+I_2$). Since there is a narrow path and next to it in parallel another narrow path, the current flows easier now than having one narrow path only. In our case **the total resistance is the half** of 1.2K (it equals 0.6K). Current I is therefore I = 3V/600Ohm = 0.005A (or 5mA) (half resistance makes double current). In chapter 2.2 we will come back to how we calculate the total resistance of resistors in parallel which may not be of the same value as in the previous case.

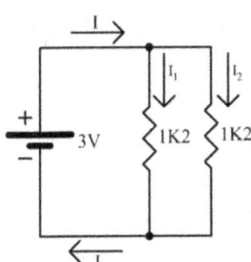

Still resistors look like they offer nothing exciting. Let's move to some switches and lights.

1.3 SWITCHES, LEDS AND SOME CONNECTION TOPOLOGIES TO PLAY WITH

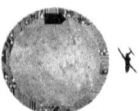

SWITCHES

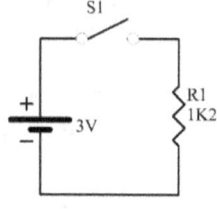

Most of you should know, or figure out what a switch is. It gets in the path of the current flow and permits or disrupts the flow, as its symbol very clearly suggests (the S1 on the left). Alternatively we may say that **it opens / closes a circuit**. There are many forms of switches, to which we will come later.

ON-OFF SWITCHES

The form described earlier has two poles (pins) and two states, ON and OFF. There are two usual kinds of switches like this, the ones that stay at the state set and the ones that are at the one state (ON usually) as long as we press them and return to the other state when we release them. Those are the buttons, their symbol is like this: ─o o─ Their electrical function is the same.

Following are some real ON-OFF switches. Their size has to do with the maximum current they can handle and with how big we want them to be (e.g. in industrial environments we may need a very big emergency stop button for easy access to it, regardless if very small current may be flowing through that switch. Same for elevator buttons for example. In a miniature circuit we may choose one of 3X3mm dimensions)

Medium current (usually around 1-3A) lever switch

Higher current (around 5A-15A) rocker switch

Slide switch (0.3A – 4A usually)

Dip switch, that is a package of many (1 to 16) switches, each holding very low current (around 50mA only)

A jumper that is placed on two adjacent pins and keeps them shorted as long as it is there

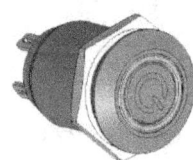
A button, a big one

A medium to big size tactile button. Tactiles are buttons which when pressed do a "click". Examples: buttons of computer mouse, volume buttons on smartphones

Other common tactile buttons. Note: if they have more than 2 pins some of those are internally connected and act as one (e.g. a 4 pins has two pairs internally connected and acts as 2 pins)

A membrane keypad consisted of many buttons on a flexible and adhesive material about 1mm thick with custom printing on its surface

ON-ON SWITCHES AND OTHER KINDS

Following is a list of the most common types of switches. Their schematic symbol is self-explanatory of their function. It also indicates how many poles (pins) each kind of switch has. The concept introduced here is that of a **selector**, rather than only a breaker/maker. The simplest of those is the one named ON-ON or SPDT symbolized as on the left with 3 pins. It has two posistions (states). At one of those states it connects pin #2 to

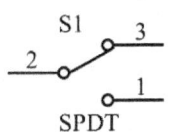

pin #3 and at the other state it connects pin #2 to pin #1. Note that if we leave pin #3 unconnected, it behaves like the ON-OFF switch. Lets see the most common of switches kinds and how they are named for finding them in the market.

Symbol	Examples	Names
	See all the above	SPST (single pole single throw) or ON-OFF
		SPDT (single pole double throw) or ON-ON or 1P2T
		DPST (double pole single throw) that contains 2 ON-OFF mechanically coupled
		DPDT (double pole double throw) or 2P2T that contains 2 ON-ON switches mechanically coupled

There are many other kinds of switches such as 6 positions rotating switches e.g. 1P6T but we keep with the practical and frequently used components only.

When we select a switch, we take into account, the fitting to our board or panel, the looks, the cost and most importantly if it withholds **the maximum current that may pass through the switch**, e.g. if the max current is 0.8A we should chose a switch with specification of 1A or more.

LEDs

At last we are in one of the most exciting and beautiful components. The LEDs with that deep, **almost magically colored light**. Starting doing a little of categorizing stuff, we may separate them in powerful, usually white colored LEDs for lighting up a space (like a 230V ceiling lamp consisted of LEDs, LEDs flashlight lamps etc.) and to indicating LEDs used not to light their surrounding environment but rather themselves, usually with a color, to indicate something. We will deal with the second category.

So... what does LED mean? **L.E.D. stands for Light Emitting Diode** and is pronounced "el ee d". What is a diode? We will see this more on the next chapters, it allows current to flow in one direction only and is symbolized with: ─▶|─ LEDs are symbolized with: ─▶|─

The most distinctive feature of LEDs compared to hot filament lamps is that LEDs do not heat anything to thousands of degrees Celsius to produce light but rather they use a quantum physics phenomenon of electric field and electrons (electrons have only discrete energy levels, light's photons have discrete energy levels also) to produce **single frequency** light (actually spread around a single frequency but only a little bit). Light is an electromagnetic wave, single *frequency* is equal to single *wavelength*, is equal to single *color* of the infinite colors we see on the rainbow, or more technically, a single *narrow band of the visible spectrum*. That makes their nice looking light and the need of very low power per light they produce. The indicating LEDs practically do not heat up.

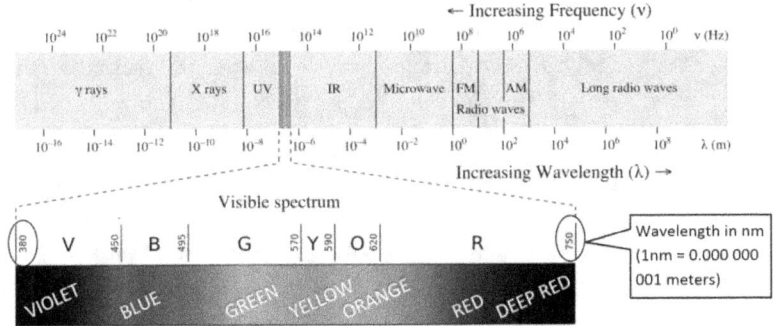

LEDs usually come in red, orange, yellow, green, blue and white color. A special case is white color LEDs which internally consist of a red, a green and a blue LED. There are also **RGB** LEDs which are again a Red, a Green and a Blue packed on the same case but providing each LED's pins (leads) to drive them separately. That way we may produce any color. Besides LEDs emitting visible light, there are also LEDs that emit light just outside of the visible spectrum of light, below the red color frequency (infrared or **IR**) and rarely just above violet's frequency (ultraviolet). Infrared LEDs produce strong and invisible light and are used in TV remote controls, proximity sensors, optical mice and cameras that can see at the dark. The most classic LEDs are the cylindrical shapes of **5mm** and of **3mm diameter** with a lens on the top, made either from clear (transparent) material or of a colored material that diffuses light. Their usual cost is around 5 cents each for ordering some dozens of them.

Let's go to the **Voltage** and **Current** issues in order to light up an LED. LEDs need around 1.5 to 2.8V to operate (each model (manufacturer part number) has its own). Below that voltage no light comes out, but applying just about 0.5V more voltage than that is catastrophic, the LED gets burned. The current an LED needs is usually 20mA maximum (to shine fully) and any current less than that flowing through it produces proportionally less light (e.g. 1mA will produce the 1/20th of the full light potential). It is suffice to say now that every LED **needs a resistor in series** to set the LED current to the value we want it to be and protect it from rising the voltage across its pins beyond the maximum. In chapter 2.5 we will come back to this to see how we calculate the value of this resistor (that is usually in the region of 30 to 300 Ohms). LEDs, as diodes, have only one direction for current flow. The two pins of a diode or an LED are called Anode and Cathode, current flows from the Anode to the Cathode. The voltage applied to them must have one "direction" only, higher at the Anode and lower at the Cathode, otherwise no current will flow. The "direction" of voltage (which one of two points is + and which one is -) is called **voltage polarity** and has to be respected.

LET'S PLAY WITH SOME CIRCUITS

First of all, we will explain another concept for circuits that is critical. It is that of the "**node**". What a node is: It is **all pins that are connected together** by one or more wires. In circuit's schematics, wires are ideal wires, meaning they have no unintentional resistance. If any practically meaningful unintentional resistance was there (wire's resistor) it should be displayed with the resistor's symbol. On the left we have a circuit with 3 nodes. No matter how long wires we use in our schematic e.g. making the simplest path or making a maze with the wires of one node, it is still the same one node, all pins it connects are connected the same way, **as if they are connected on a single point**. The reason this happens is that a zero resistance conductor **has the same voltage across any of its points** since (considering it is a resistor) V=I*R where R=0 so V=0 regardless the current's value. For example at the previous circuit, voltage is the same for points A, B, C and D. If we touch the one probe of our voltmeter (multimeter measuring voltage) at A and the other at D, the voltage measured will be zero. If we touch one probe to any point at node 3 and the other probe to any point of node 1, the voltage measured is going to be always that of the battery.

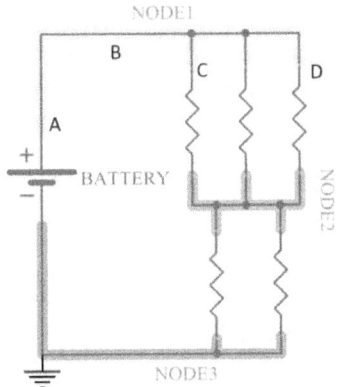

A note: since it is tedious to say every time "voltage across point X and point Y" we use to name the negative pole of the battery / supply "**ground**" and refer always to this as the negative point. That way we may say "Node1 has voltage X Volts" implying across ground that is node 3. Ground symbol is: ⏚

About the current there is a different story. That may branch at a node to separate "rivers" or be merged, as described in the chapter 1.1.

Let's meet our first real circuits! We will use them all for understanding connection concepts, from simple to more and more complex. Their full understanding is very important.

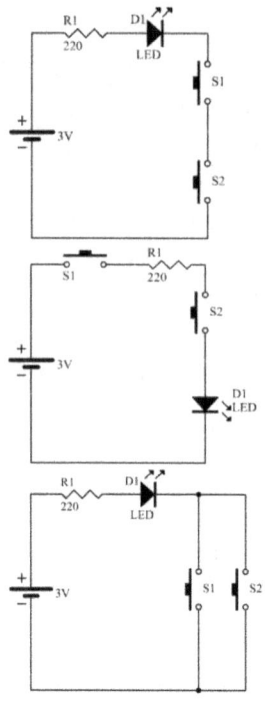

1. Both buttons (S1 and S2) have to be pressed (concurrently) in order to light the LED

2. The same applies for any "sequence" of placing components which are in series

3. Pressing ANY of the buttons, the LED will shine. Same if both buttons are simultaneously pressed.

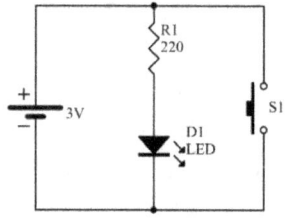

4. A very bad idea! In an ideal world of components and wires, pressing the button will provide the current a path that has zero resistance. The less the resistance, the higher the current, so infinite current (I = 3V/ 0Ohms) will be a disaster. Sorting a battery is bad for its health and perhaps for the health of the switch (max. current specification) and maybe for seeing thin wires melting after burning their plastic insulation. If the voltage source is a power supply it may be damaged at prolonged time or it may use a protection (e.g. resettable fuse) to cut off the current itself. That's about **sort-circuiting** a voltage supply… (connecting the voltage poles with very low resistance)

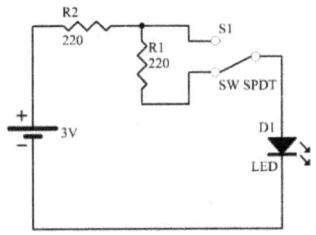

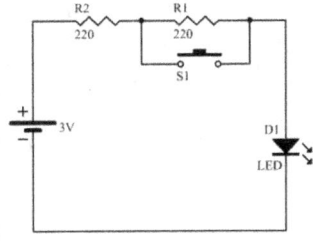

4. When S1 is "down" as seen on the schematic, the total resistance is R1+R2 = 440 Ohms. When S1 is "up" the total resistance is R2 = 220 Ohms. So in the "up" position the LED's current should be the double and the illumination of the LED will be double than in the "down" position.

5. Pressing the button (S1) will provide an alternative path for the current around R1 with zero resistance, so the total resistance in series to the LED will be 220 Ohms (=R2+0). Releasing the button will make a total resistance in series to the LED equal to R1+R2 = 440 Ohms. So… Button released, the LED shines, button pressed, LED shines with twice the illumination.

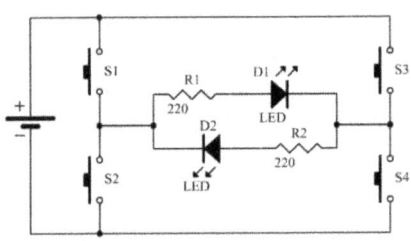

6. When both S1 and S4 are pressed D1 shines. When both S2 and S3 are pressed D2 shines. You should not press simultaneously S1 and S2 or S3 and S4 of course

Those have been just some introductory concepts with broader applications to most of electronics. Playful and practical enough themselves. But it is surely limited what we can think of having only LEDs, resistors and switches. Let's open some more chests with toys to play with…

1.4 Passive components fly over

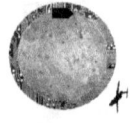

A classification used for components is to divide them between passive and active. Passive components usually have 2 pins and they do always the same action to Voltage and Current depending on the value of those two or to how they change over time. For example, the resistor regulates the current that passes through it according to the voltage across it. A 1K resistor, any time it has 2V across it, it will have 2mA of current flowing through it, every second, every hour, every year.

Active components on the other hand allow controlling. What they do depends on a "controlling signal", so they are not so "dump". Imagine as an example, a switch with an input pin that turns ON or OFF depending on whether that pin has non-zero or zero voltage. Imagine a whole computer inside a single component (chip). We will come to them on the next chapter.

Passive components are everywhere in electronics circuits. Let's learn about them starting from the most frequently used going gradually to the less and less used.

Capacitors

After the resistors, the most frequently met component is the Capacitor. Unfortunately its behavior is not as simple as the resistor's but this world is not simple anyway...

The capacitor's symbol is: ─┤├─ and the letter it is used for abbreviation is (you guessed right) "C". So... what does the capacitor do? We will describe it in vague terms and in approximation in this chapter.

Capacitors store temporarily energy (electrical), hence their name suggests how much energy they can store. The physical effect is called **capacity**. It's really like the capacity of a car's reservoir in fuel, but instead of gasoline, it is literally electrons. Think of them as tiny rechargeable batteries, but really small. If a battery can store

X energy a capacitor of the same size can store about X/100,000 energy! (Exceptions are "ultra-capacitors", which you will almost never find in a circuit, with record X/100). Moreover capacitors are usually tiny in size, the energy they can store usually lasts for milliseconds or microseconds, yet they are very much useful, practically **in every 3 - 5 components, one is a capacitor**.

Like resistance is measured in Ohms, capacity is measured in **Farads (F)** (coined after the English physicist Michael Faraday in the 1860's). An average capacitor found in an electronic board is around 1μF (microFarad = 1 millionth of a Farad – see Appendix 3.1), ranging in practice from 1pF to a few thousands μF. But they are not characterized by the capacity only. They are also characterized by the maximum voltage they can handle in Volts. Let's see how they look like.

A capacitor is formed by pacing two metal plates close to each other, keeping them electrically insulated.

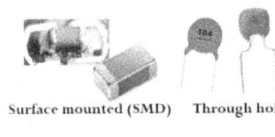
Surface mounted (SMD)   Through hole

Small ones made from ceramic materials ranging usually from 10pF to 10uF tolerating form 10V to 50V. They cost around one cent.

The biggest ones are called Electrolytic Capacitors (they use electrolysis). They range from 10uF to about 10,000uF. Their problem is (besides other problems) that they have polarity, one pin (lead) has to be always more positive than the other otherwise they brake (with a little explosion sometimes!). Their symbol is:

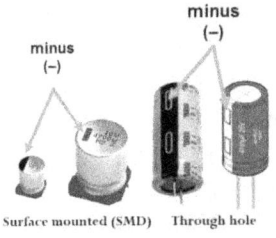
Surface mounted (SMD)   Through hole

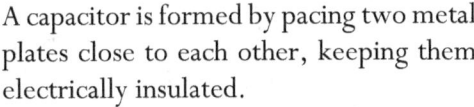

Let's take a glimpse of what capacitors do. Imagine them as a rechargeable battery with the following differences: They charge – recharge quite fast, even in a few nanoseconds (0.000 000 0001 seconds), they can be charged and discharged infinite times without any effect in their service life and as said earlier the charge they can store is about a million times less.

1.4 Passive components fly over

Capacitors store electric charge (electrons) inside them. Like batteries (but using a way simpler mechanism than chemical reactions) when current flows through them, the "charging level" increases proportionally to the voltage across their pins.

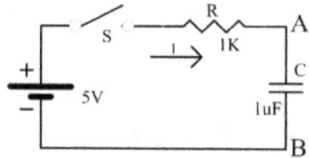

Let's see how we may **charge** a capacitor (tip: no wall plug charger is needed):

Let's assume switch S open and capacitor C not charged initially. The Voltage across its pins should be $V_{AB} = 0V$. Let's close the switch and start counting the time (time in our stopwatch when we close it is 0 seconds). Initially the capacitor will have $V_{AB} = 0V$, so the resistor R will have 5V across its pins. The current I at time t=0 will be equal to $I = V/R = 5V/1000 Ohm = 0.005A = 5mA$. On the table on the left is how the Voltage across the capacitor (V_{AB}) and the flowing current will evolve over time. We will come to analytical formulas for doing those calculations in chapter 2.5. So we see that our "battery" charged to about 60% in 0.001 seconds and to 100% in about 0.01 seconds. After that time our circuit is like the circuit on the left. No current is flowing (having zero Volts across R) forever until we start somehow drawing current from our "battery".

Time (sec)	V_{AB} (V)	I (mA)
0	0	5
0.001	3.161	1.839
0.002	4.323	0.677
0.003	4.751	0.249
0.004	4.908	0.092
0.005	4.966	0.034
0.006	4.988	0.012
0.007	4.995	0.005
0.008	4.998	0.002
0.009	4.999	0.001
0.01	5.000	0.000
0.011	5.000	0.000
0.012	5.000	0.000

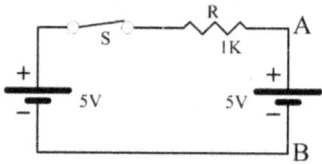

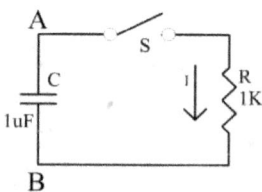

Let's use our charged battery now to power something (**discharging** our capacitor). Let's put a resistor for this. Assume C is fully charged. Closing the switch again at the time we start our stopwatch we get the mirrored results and yes, our "battery" discharges in 0.01 seconds. Increasing the resistor the charging and the discharging times increase, increasing the capacity they also increase (proportionally: doubling of any of them doubles the time). Having this quantified example makes us more familiar to the behavior of a capacitor, but, is this any useful? How can the capacitors be the 1/3th of all electronic components?

In about the 95% of the cases a capacitor is used in practical circuits, is to provide a **"charge reservoir"**. All such uses **stabilize the voltage** that supplies something i.e. the voltage will never fluctuate a lot. One use is to stabilize voltage fluctuations happening by very sudden current fluctuations. Digital circuits require sudden current supply of tens milliamps for sort times. How short? Nano seconds only. A close-by capacitor will act as a local battery that will be discharged by a negligible percentage while it provides such sudden and sort period current demands. Another use of a "reservoir" is to stabilize a fluctuating voltage. Imagine a voltage used to supply stuff (like our previous batteries) that ripples (fluctuates a little). Let's assume that the supply voltage fluctuates from 4.9V to 5.1V every 1usec, or 1milion times per second. Such frequencies are classic in switching power supplies. At such time scales, the discharging of this capacitor with this resistor combination (that is also 1K / 1uF as in our previous example), evolves over time as we see on the left (notice time is now usecs). We see that in 1usec period, V_{AB} fluctuates so little in

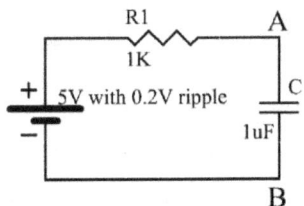

Time (usec)	V_{AB} (V)
0.0	5.0000
0.2	4.9990
0.4	4.9980
0.6	4.9970
0.8	4.9960
1.0	**4.9950**
1.2	4.9940

comparison to the supply voltage fluctuation. This is because our capacitor has so much charge reserved (like a big battery) that been discharged by so little, it holds its voltage practically steady.

Those had been simplified (with real numbers) illustrations in order to see some of the capacitors applications without their details, but to feel how they work.

DIODES

Don't worry, diodes are not as complex as capacitors. They are actually a lot simpler that almost all components.

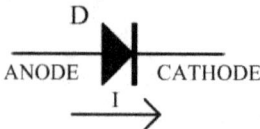

They allow current to pass only in one direction, yes, their symbol is "D" but the names of the two pins it has are bizarre, they actually come from the vacuum tubes electronics era (old screens name "CRT" is for "Cathode Ray Tube" for example). This is how they behave according to the voltage polarity across their pins:

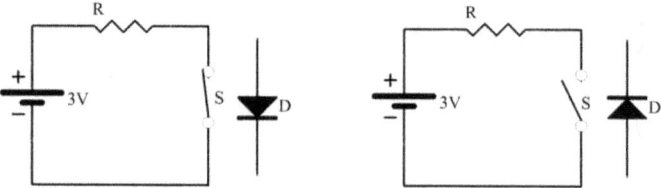

The real world diodes unfortunately have not precisely this ideal behavior, we will see their deficiencies and diode kinds in chapter 2.5. It is now suffice to say that when they are "forward biased" as in the left of the above circuits, they "drop" the voltage by a bit (about half a Volt) as if they are containing an internal battery like the circuit on the left.

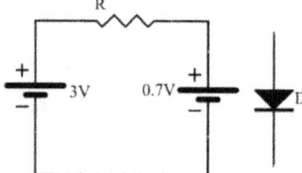

Let's see some:

They usually carry a mark of a dash for the cathode pin (just like the dash in the schematic symbol). They are mostly characterized by the maximum current they can tolerate since that makes them heat up (more or less their size is following that).

COILS

Coils are the magic of electromagnetism. Turn any wire like a 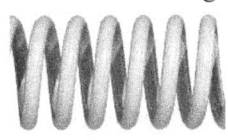 "spring" shape and you have a coil! Add a piece of magnetizing material in it (like iron) and you have a coil with a core that has more "coil" activity. Here are how some pre-fabricated coils look like, which you can buy ready to use:

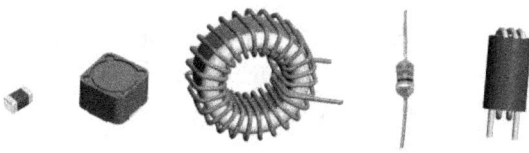

Coils are less frequently used (expect one every 30 components) and that's fortunate because their way of working is quite hard to understand. We will pass them quickly.

The physical effect they do is called **inductance** and is measured in **Henry (H)** after Mr. Joseph Henry in 1870's. Usual coils range in a few uH (micro Henry). Another specification is the maximum current they can tolerate.

The magically complex work of inductance is this: It reacts to changes of current flowing through the coil by generating a voltage across its pins (leads) that will make it hard for the current to change. It also behaves like a battery in the concept of "charging" and "discharging" but instead of how much charge it has stored, it is about how much current (Amperes) are flowing through it.

Actually coils (inductance in general) store electrical energy in an internal magnetic field they generate that is stronger the more current is flowing through their wire. The practical uses of the coils are:

- o To stabilize current when that fluctuates unintentionally or to cut off any signals alternating in high frequencies.
- o Mostly: to do the magic to raise, lower or invert a voltage without losing energy in heating itself. How? Let's see about raising the voltage. When the current in a coil drops, it raises the voltage across it in order to make it hard for the current to drop and that voltage rising can be exploited in the so called "step-up DC/DC" converters (visited in chapter 2.6).

1.5 ACTIVE COMPONENTS AND ICS FLY OVER

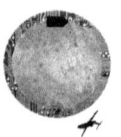

The more we travel the electronics planet, the more useful stuff we find. Active components are - at last - components which can "control" stuff or do operations from the very least to very high intelligence.

They divide in discrete parts and integrated circuits (ICs) or chips. The latest are comprised internally by the first. So let's begin with the discrete components which are the ingredients of the ICs.

FETs & MOSFETS

MOSFETs are the most important control element of all modern electronics. Computers are literally comprised of MOSFETs, quite a lot though. They are the "bricks" of the digital electronics. What are they?

Metal Oxide Semiconductor Field Effect Transistors (MOS-FETs) are switches (ON-OFF kind) made of Silicon (Si) with a control 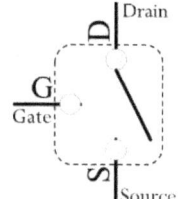 input. The voltage at the control input sets the switch ON or OFF. The switch may open or close circuits that set the voltage of other control inputs (implementing digital "logic") or they may just "drive" high current demanding loads such as big LEDs motors etc. The control pin is called "Gate", the switch's poles are called "Source" and "Drain". The important control voltage is the voltage across the Gate and the Source pins (V_{GS}). If that is lower than a value (around 2V) the MOSFET is an OFF-state-switch doing nothing. When V_{GS} is higher, Drain is connected with the Source. The most practical application of using MOSFETs is to turn ON or OFF powerful things like motors, big lamps etc. which require a current flow from hundreds milliamps to many amps using very small current to create this V_{GS} voltage. Actually this current is negligible, it is practically zero. We use a small power source that is controlled by "logic" as we say in digital electronics to control a big power source, just like we control the motor of our car with a negligible force in our foot on the gas pedal.

MOSFETS are divided in two main categories, N-channel and P-channel. N-channel require the Gate voltage to be more positive than the Source voltage (usually their Source is connected to the negative pole of the power supply / battery that is the ground), P-channel do the opposite. We will dive into the details of them later, for the moment, following are their schematic symbols and some pictures.

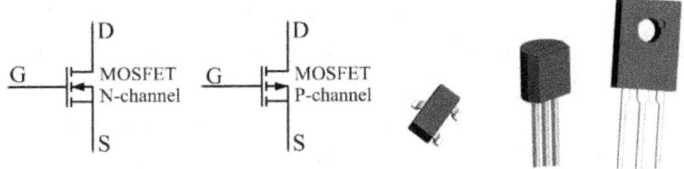

The main specifications for choosing the right MOSFET are the maximum current its switch can hold and the V_{GS} Voltage threshold required to turn on.

Transistors (BJTs)

In December 1947 at the Bell Telephone Laboratories the course of electronics changed from Vacuum Tubes to tiny parts made of solid material (silicon mixed with some other) called solid state electronics. Those had been the transistors. (More accurately Bipolar Junction Transistors, since FETs fall in the transistors category also, but it is used to call them just "transistors").

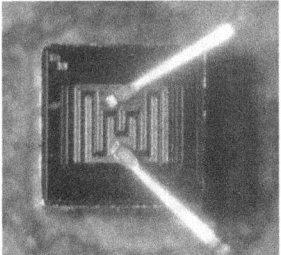

Transistors do what FETs do but they do not act according to a control voltage but rather they act according to a control current. Their 3 pins are called with the strange names: Emitter, Collector and Base. The control current enters the Base and exits on the Emitter. It must be over some threshold in milliamps (around $1/20^{th}$ the current

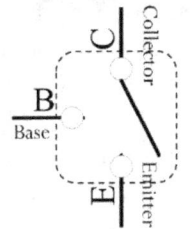

that passes through the switch) in order to turn the switch ON. There are two types: NPN (Negative Positive Negative) and PNP

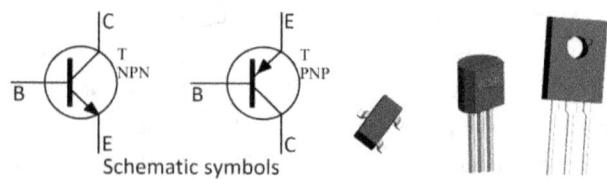
Schematic symbols

The most used ones are the NPN type. MOSFETs nowadays are more common in electronics than transistors (BJT) since they are more ideal switches than BJTs as they practically do not require current to keep the switch ON. We couldn't leave not mentioning the transistors though in a book about electronics. What actually happens in the silicone that makes them work (also the FETs) is in the field of "solid state physics". We avoid analyzing stuff so little that it cannot be understood. It's a long story about "electrons mobility" and electric fields in the worlds of the atoms. It is the nucleus, the essence of electronics, but it should take some chapters to be fully presented. Its magic physics.

A final note is, both FETs and BJTs do not act as switches only but also as amplifiers when they barely switch ON. This is not used in practical electronics nowadays, with negligible cost we buy amplifier chips doing that job, in the 99% of the cases a lot greater than single transistors.

Integrated Circuits (ICs)

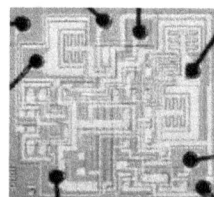

Let's meet our first Integrated Circuits (ICs or chips). Since 60s more than one transistor is fit inside a small part of silicon. Resistors and capacitors are also fitted. At the later years we count over a billion transistors in a silicon piece (called silicon die) of sizes about one inch by one inch. The technique of making those is the grail of our digital civilization. Each element of an IC is not made individually. Masks are used, having all of them as drawn patterns, light shines with the silicon die underneath, photochemical reactions take place and all the chip is made after doing that sometimes with different masks and chemicals. No matter how many transistors are drawn on those masks, like making a photocopy, if the paper we are copying contains one line or a thousand shapes, the labor and the process time are the same. The silicon die is packed in a case like the one on the left connected internally by very thin wires to the outside pins. There is a diverse variety of IC cases or packages. Some common are:

SMD or SMT (Surface Mount Technology)

THT (Though Hole Technology)

ICs USEFUL FAMILIES

ICs do almost everything voltage and current can do in our wildest imagination. The measure of the skills of an electronics designer at providing most practical (simpler) and economical solutions is mostly about her/his knowledge and experience on which chip on the market will do the job best. At the time of writing this book, the (perhaps biggest) components distributor (e-shop if you like) digikey.com lists 713190 IC models, 42085 of those sited as "new products". That is not to scare but to make the filling of living in Alice's wonderland where we can point our mouse on this link and click it. Let's see the most interesting ICs categories from the • Embedded - Microcontrollers (74052 items) simpler to more and more complex:

- Regulators: They manage to convert a voltage from one value (input) to another (output) that is well-defined and steady, even if the input voltage varies. Those supply well our circuits.
- Amplifiers and comparators: They amplify or compare analog signals (any voltage value within reasonable range)
- Load switches – H-bridges: They provide easy control and high current outputs to current demanding devices like motors, big LEDs etc.
- Digital logic ICs: There is a diverse variety of simpler logic ICs like logic "gates", serial to parallel conversion, pulse output of special timing etc.
- Sensor ICs: Sensors measure a physical quantity (like temperature) by converting it to electric signal like a voltage we can measure, or to a digital information we can use. Some sensors are integrated inside ICs providing advanced functionality in many fields of measurements.
- Microcontrollers (MCUs): The core of all modern electronics called **"embedded electronics"**. They embed a computer (CPU and memory) and many "peripherals" such as communication interfaces, analog to digital signal converters, timers and others. They are the core of this book. They are coming after 3 chapters. Hung on…
- FPGAs: Magic chips where you can define (program) how they will internally interconnect to form any digital circuit you want. But they are complex and rarely needed in practical electronics, let's catch them later when we do special NASA missions ☺

1.6 COMPONENTS CONNECTING TECHNIQUES

Let's take our first steps in the path that goes from "learn" to "make". How do we really make circuits?

For start we must consider: a) there is not one way and b) there are a lot of levels between a circuit made in our garage for fun and a circuit recently produced by Apple (except "Apple I" also produced in their garage ☺). We are focusing in practical electronics. Let us now take a view of the simple, easy and playful ways to build, or "make" our circuits, as the "maker" term is the recent trend. We will present higher level methods (more reliable, hi-tech and professional) in chapter 2.4.

BREADBOARDS

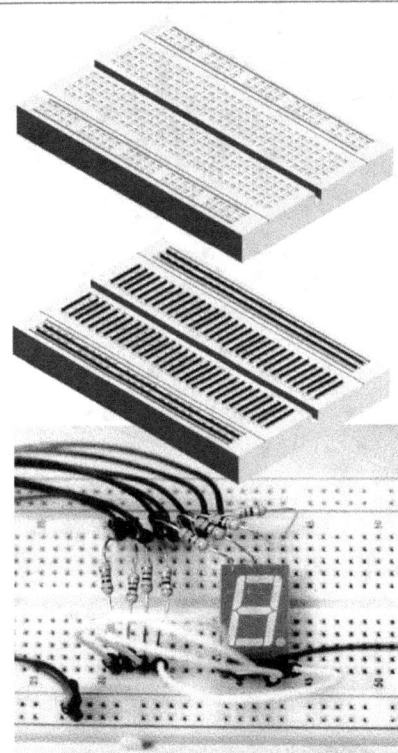

The easiest and dirtiest way to make a circuit fast is to use a breadboard, THT (through hole) components and some wires. Breadboards have a really bizarre name and they are like the one on the left. They cost around 2$ and have quite a lot of holes. The way they work is quite genius. If you insert a wire or a pin in any of those holes, it is held in place by metallic sideways springs which also connect it electrically to each of the neighbor pins we see in the left pattern. Finally projects made on a breadboard tent to look like that on the left. The best thing about them is that we do not harm the components (no soldering, not even cutting their wires) and they can be easily removed and re-used. Another is the record quick time to make them. On the other side, they are bulky, not very reliable (oxidized pins, loose spring conducts or not fully inserted pins are causes) and become like a spaghetti soon enough. Generally what you make on a breadboard is specified to work only on the table where you made it. Let's see some examples from schematics to breadboard implementations to understand them better.

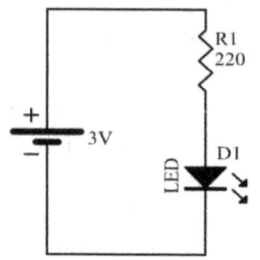

The program used to make the images is the great open source software called Fritzing (fritzing.org). I am grateful to this team for that.

Let's start with the circuit on the left. An implementation can be the following one:

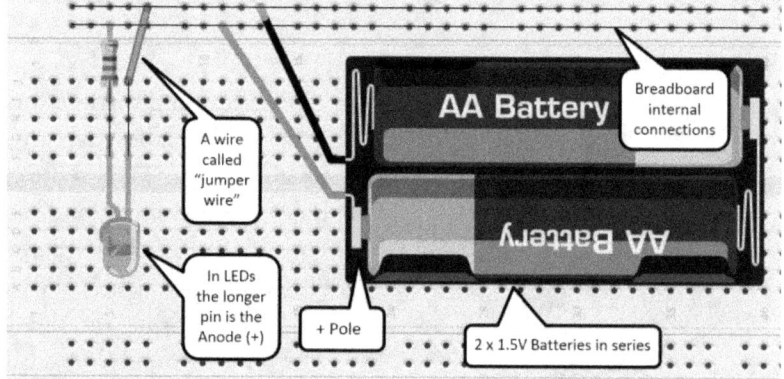

Of course many other combinations can make the same exactly circuit. A handy feature of breadboards is that their design allows to place ICs like the illustration on the left and populate around them easily many components.

But if you have ever seen the least of an electronics laboratory (lab) you have surely seen a soldering iron in action. Soldering is the way to make more reliable and compact circuits.

PCBs

When we say "electronic board" we mean Printed Circuit Board (PCB) with electronic components mounted on it. This is the top way we may construct our circuits. The components are all soldered as we will see in the next paragraph. PCBs are a flat, insulating and heat resistant material, with patterns made of copper foil on its surfaces that connect component pins together. So instead of wires we have a rigid body with conductive tracks running in complex patterns connecting point to point. We will present them analytically in chapter 2.4

Soldering

To solder you need solder! Solder usually comes in wire form made of a metal that melts easily (at about 200°C = 400°F). Such temperature is found on all flames (usual flames are 4 times hotter), on an iron for cloths, on a cooking stove. It's a far lower temperature than the thousands of degrees on electric arc welding or oxygen welding, yet it welds (solders) some metals like copper and nickel very well. So, if a soldering iron tip touches your hand

you might and you might not get a minimum burn. Soldering Irons are devices which heat a tip at the right temperature to melt solder wire and their tip is at the right geometry to interact with electronic components of some size. It requires some exercise to create some skills on it, but it takes only a few hours of exercising to become skillful enough. The process of soldering is terribly simple. The solder wire is solid when it's colder than its melting point (room temperature) and liquid in temperature over its melting point. While it is liquid it flows like a small drop of

water on the tip of the soldering iron. We move that drop where we should in order to join together wires, pins, copper pads on PCBs etc. (we will come later on those), remove the tip, wait a fraction of a second to cool down and that drop is solid, frozen at the place we left it, holding the parts together. Any smoke coming out is

fumes from a substance added in the solder wire called "flux" that helps flowing when it is melted. It should be avoided to inhale, it is not too dangerously toxic though. We will fill all the important details about this technique in chapter 2.4.

1.7 REGULATORS – A FIRST GLIMPSE TO ICs

Almost every useful circuit contains at least one regulator. In this chapter we will take a quick look at regulators and exploit that opportunity to introduce more Electronics Engineering language (terminology) and how ICs specifications look like in their "user manual" called datasheet.

What do regulators do?

Our circuits *are supplied* with voltage that needs to be steady (e.g. 3.3V all time) and be capable to provide however much current our circuit needs (*consumes*), keeping always the *supply voltage* steady. Even a big battery that has no problem to provide the circuit's current needs cannot maintain a steady voltage as it is discharged over time and its voltage gradually falls.

Regulators provide a steady voltage output that can supply our circuit and make us feel sure that this voltage will always be inside ±1% of the specified value if some specifications are respected. Guarantying that a circuit is supplied with a well specified voltage is crucial since most ICs need to be under specific voltage range and since the behavior of the whole system should be the same regardless e.g. the charging percentage of a battery. To start with the problem (in engineering terms, "problem" is a good word) *power supply sources* either have rather big changes of their voltage over longer times (a battery charges – discharges, a photovoltaic cell is shined more or less), uncertainty of the voltage (circuit designed to accept e.g. from 5V to 12V supply input) or *ripple* in the voltage, that is, changes usually periodic of 0.02 sec to 0.1usec *period*.

All regulators produce a steady *output voltage* and provide *output current* from 0 up to a maximum value (since current flow heats them up), according to how much current our circuit needs or "*draws*". Each has at least one input, one output and a *ground (GND)* pin (connects to circuit's ground) like the schematic symbol on the left. Note that ICs schematic symbols are arbitrary, they only need to show the *pins* of the IC they represent. In the pictures of ICs in the previous chapter, we saw that ICs have *pins*, from 3 to more than 100. They also have *"packages"* SMD or THT type. Pins are 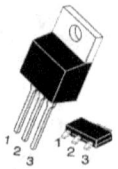 usually named by numbers starting from 1. The number given to each pin is described with pictures on the real package of the chip, like on the left. We do not want to connect wrong pins when we go from schematic to "making".

LINEAR REGULATORS

The simplest form of regulators are the *"linear regulators"*, most of which - though not all - are called *LDOs - Low Drop Out* regulators. Most are "fixed voltage output", that is, if you buy a fixed output regulator of 5V, changing the output voltage requires to change the regulator to another with different fixed output voltage. The *input voltage* in linear regulators has to be higher than the output voltage (but in a specified min – max region). Any input voltage in that region, will produce the same (fixed) and steady output voltage. Linear regulators tend to heat up easily by the current they deliver. Other kinds of regulators are the switching kind which use a coil. They may produce higher output than input and heat up a lot less, so be capable to deliver higher current output, but they are more complex.

INTRODUCTION TO ICs &
COMPONENTS SPECIFICATIONS

In *semiconductors* industry there are many manufacturers, other more, other less famous (all electronics made of Silicon are semiconductors, just because they may be conductive like a wire, or they may not be, according to conditions). There is no way to convey practical electronics engineering knowledge without referring to those. The author of this book is affiliated to none. You will see a lot manufacturers firms and a lot of products mentioned throughout this book but they are random, chosen according to subjective views and are surely not the only ones to do a job or solve a problem. Let us freely talk about any product from the hundreds thousands out there without any taboos of advertising or "burying" others. Same applies for any incorporation mentioned.

If we speak about any IC, we must refer to how to buy it and get in your hands that same one. All have a manufacturer and a **manufacturer part number**. The latest defines them fully and is unique. To search about ICs the most recommended way is to use the big distributors e-shops where you have in one place all major manufacturers' products that fall into specifications filters or keywords you provide. Finding your best IC does not mean you will buy it from there. Here are a few major such sites:

Name	Web site	Location	Subjective description
Digikey	digikey.com	USA	Best web site regarding information, plethora and navigation speed
Mouser	mouser.com	USA – Europe dispatching	Almost same as Digikey with a few less products. Ships from Italy for European customers
Farnell	farnell.com	UK	Ultimate low shipping cost, ultimate delivery speed in Europe, less products
LCSC	lcsc.com	China	Decent DHL shipping cost, includes a plethora of Chinese manufacturers and products - yet has less variety than the previous distributors but amazingly cheaper prices
Octopart	octopart.com	-	This is a multi-distributor search engine. You enter a part number and see in all big distributors the price and the stock

To choose an IC for a job first we see the important specifications (*specs*) on those sites and if we like those, its price and its availability, we download and read its **datasheet**.

A datasheet tells all the truth about the component. It leaves no untold information to the most possible. They are 100% in technical language requiring the reader to have <u>basic</u> background in general electronics and in similar products. They may be from 2 pages to 2000 depending on the product. They are not for ICs only, every electronics component, even a resistor has its manufacturing part number and a datasheet.

We are in the real world. Almost nothing works ideally. Even resistors have value tolerance, maximum power dissipation, side effects like a little inductance and other specifications referring to environmental conditions they can survive in. They also have geometric dimensions, handling recommendations etc. We will try in this book to make you understand the most practical specifications of datasheets and the descriptions the make. The path to learn more of applied electronics, after reaching a basic level, is to read components datasheets.

BACK TO THE REGULATORS: LD1117V33C

Let's take one out of the thousands of regulators, which is common, cheap and broadly used. It's the LD1117V33C part number from ST Microelectronics. A practical way to get its' datasheet is to search for LD1117V33C in Digikey and be right on it. As a bonus we learn the price that is about 0.5$ each and a summary of its specs. No need to do it now, we will take a feeling of what is written in it and of how we will use our regulator in the following paragraphs.

It is a common practice datasheets to be written for a series of products as in our case for LD1117 series. That series includes different package sizes and different voltage output part numbers. Let's see some datasheet's highlights (material taken from June 2019 document version)

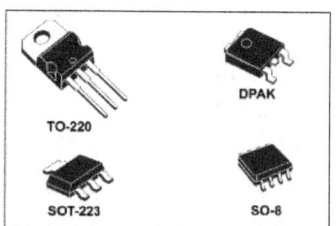

Description

The LD1117 is a low drop voltage regulator able to provide up to 800 mA of output current, available even in adjustable version (V_{REF} = 1.25 V). Concerning fixed versions, are offered the following output voltages: 1.2 V, 1.8 V, 2.5 V, 2.85 V, 3.3 V and 5.0 V. The device is supplied in: SOT-223, DPAK, SO-8 and TO-220. The SOT-223 and DPAK surface mount packages optimize the thermal characteristics even offering a relevant space saving effect. High efficiency is assured by NPN pass transistor. In fact in this case, unlike than PNP one, the quiescent current flows mostly into the load. Only a very common 10 µF minimum capacitor is needed for stability. On chip trimming allows the regulator to reach a very tight output voltage tolerance, within ± 1 % at 25 °C. The adjustable LD1117 is pin to pin compatible with the other standard. Adjustable voltage regulators maintaining the better performances in terms of drop and tolerance.

The devices packages names are universally standardized in geometry. Any component in TO-220 has the same shape and dimensions. Do not try to follow yet the non-highlighted text. It is of lower importance anyway. In the first 7 highlighted lines "drop voltage" is how much higher the input voltage has to be from the output voltage. In adjustable regulators we can set the output voltage using two resistors. Let's see what the next group of highlighted lines tries to tell: In a next page (p.8) we see:

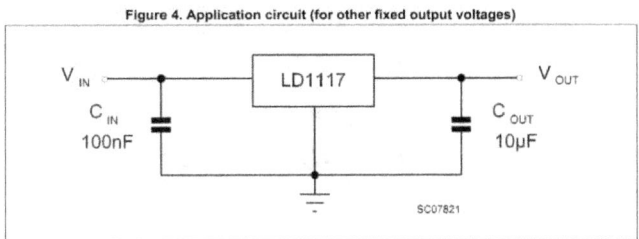

Figure 4. Application circuit (for other fixed output voltages)

That is how almost all regulators are used. They require 2 capacitors placed close to them for charge reservoirs, one at the input and one at the output. Other specification highlights are:

Table 1. Absolute maximum ratings

Symbol	Parameter		Value	Unit
V_{IN} (1)	DC input voltage		15	V
P_{TOT}	Power dissipation		12	W
T_{STG}	Storage temperature range		-40 to +150	°C
T_{OP}	Operating junction temperature range	for C version	-40 to +125	°C
		for standard version	0 to +125	°C

1. Absolute maximum rating of V_{IN} = 18 V, when I_{OUT} is lower than 20 mA.

Absolute maximum ratings inform us when and how we burn our component, e.g. input voltage higher than 15V will burn it. We should always look twice at all of them.

Table 25. Order codes

Packages					Output voltages
SOT-223	SO-8	DPAK (Tape and reel)	TO-220	TO-220 (Dual Gauge)	
LD1117S12TR		LD1117DT12TR			1.2 V
LD1117S12CTR		LD1117DT12CTR			1.2 V
LD1117S18TR		LD1117DT18TR	LD1117V18		1.8 V
LD1117S18CTR		LD1117DT18CTR			1.8 V
LD1117S25TR		LD1117DT25TR			2.5 V
LD1117S25CTR		LD1117DT25CTR			2.5 V
LD1117S33TR	LD1117D33TR	LD1117DT33TR	LD1117V33	LD1117V33-DG	3.3 V
				LD1117V33C-DG	3.3 V
LD1117S33CTR	LD1117D33CTR	LD1117DT33CTR	LD1117V33C		3.3 V
LD1117S50TR		LD1117DT50TR	LD1117V50	LD1117V50-DG	5 V
					5 V
LD1117S50CTR		LD1117DT50CTR	LD1117V50C		5 V
LD1117STR		LD1117DTTR	LD1117V	LD1117V-DG	ADJ from 1.25 to 15 V
					ADJ from 1.25 to 15 V
LD1117SC-R		LD1117DTC-R			ADJ from 1.25 to 15 V

Near the end of the document we find the specifics of our manufacturer part number. It is 3.3V fixed output in a TO-220 package. If we dig a little more inside the document, we find that LD1117V33 works from 0°C - 125°C, while LD1117V33C works from -40°C to125°C. This table also helps to select other part numbers according to our needs.

Last, the detailed *electrical operation characteristics*:

Refer to the test circuits, T_J = -40 to 125 °C, C_O = 10 µF, unless otherwise specified.

Table 12. Electrical characteristics of LD1117#33C

Symbol	Parameter	Test condition	Min.	Typ.	Max.	Unit
V_O	Output voltage	V_{in} = 5.3 V, I_O = 10 mA, T_J = 25 °C	3.24	3.3	3.36	V
V_O	Output voltage	I_O = 0 to 800 mA, V_{in} = 4.75 to 10 V	3.16		3.44	V
ΔV_O	Line regulation	V_{in} = 4.75 to 15 V, I_O = 0 mA		1	30	mV
ΔV_O	Load regulation	V_{in} = 4.75 V, I_O = 0 to 800 mA		1	30	mV
ΔV_O	Temperature stability			0.5		%
ΔV_O	Long term stability	1000 hrs, T_J = 125 °C		0.3		%
V_{in}	Operating input voltage	I_O = 100 mA			15	V
I_d	Quiescent current	V_{in} ≤ 15 V		5	10	mA
I_O	Output current	V_{in} = 8.3 V, T_J = 25 °C	800	950	1300	mA
eN	Output noise voltage	B = 10 Hz to 10 kHz, T_J = 25 °C		100		µV
SVR	Supply voltage rejection	I_O = 40 mA, f = 120 Hz, T_J = 25 °C V_{in} = 6.3 V, V_{ripple} = 1 V_{PP}	60	75		dB
V_d	Dropout voltage	I_O = 100 mA, T_J = 0 to 125 °C		1	1.1	V
		I_O = 500 mA, T_J = 0 to 125 °C		1.05	1.15	
		I_O = 800 mA, T_J = 0 to 125 °C		1.10	1.2	
V_d	Dropout voltage	I_O = 100 mA			1.1	V
		I_O = 500 mA			1.2	
		I_O = 800 mA			1.3	
	Thermal regulation	T_a = 25 °C, 30 ms Pulse		0.01	0.1	%/W

At this point some of those will be out of comprehension. Patience... A last note about the LD1117V33C: Since it is a well-selling IC, many other manufacturers have made compatible ones using manufacturer's part numbers like xx1117xxxx.

1.8 ONE PLUS ONE MAKES 10: THE BINARY SYSTEM AND THE DIGITAL WORLD

Computers, data, nowadays media storage, communications, even TV are termed as "digital". In this chapter we will demystify what "digital" is and how much simple and stupidly "tricky" it is. The word comes from "digit" but that is only misleading. Let's take a break of all those components we dealt with and travel a little in some ideas, in the world of digital logic.

ANALOG ELECTRICAL SIGNALS

What is electrical **signal**? Anything that carries information electrically. In 99% of cases, information is corresponded to voltage's value (the other case is current's value), such as sound, data etc. Electromagnetic waves are not signals. Electrical signals are also "control signals" such as a voltage applied on a MOSFET's gate that controls its "switching state" to ON or OFF. Even if that changes very seldom, it is information also that sets a state of something.

What is not an electrical signal: Voltages that are intended to be steady forever, such as power supply voltages and others few. So in electronics most voltages are signals.

A signal (voltage) can take any value and between two voltage's values which are very close, there are infinite values a voltage may have in between (e.g. there are infinite values a voltage may take between 0.95 and 0.0950001 Volts). If we care about how much a voltage is and even a negligible difference makes a difference to us, that's an **analog signal**. Since nature works in a way that allows practically infinite "resolution" of a voltage, all signals are really analog, but they are characterized as analog if any difference in voltage value makes a different result e.g. driving an LED with a variable voltage that is intended to set it's intensity. The more voltage the brighter the LED. Same for sound that is just converted to voltage using a microphone.

All analog signals suffer from some problems:

- Accuracy: even if the voltage value has absolute accuracy of 0.000001% it has randomness (uncertainty) in how much exactly it is. That is always the case for any accuracy requirement, tight or loose.
- Distortion: When an analog signal is passed through electronics or just through wires which are not intended to alter it, it gets altered almost always (e.g. the higher the frequency the more the attenuation)
- Noise: This is the biggest problem. Noise is random fluctuations added to the signal. Nature adds noise almost at all systems. Examples: a hissing noise on sound, specks on picture of analog TVs, scratches on vinyl discs, etc.

THE BIG TRICK: TWO STATES ONLY!

Here is the great trick. We take a signal (analog signal) and we divide the voltage value in two **discrete** areas, "high" and "low", let's say below 2V and over 2V. Let's now produce signals (voltage) **that is either 0V or 3.3V**. Accuracy, distortion and noise cannot be so terrible as to make our 0V voltage to over 2V or our 3.3V voltage to less than 2V. (2V is not such an exact threshold, it is rounded for our example). Taking into consideration only if the voltage is "high" or "low" we skip all analog signals problems. We have a method that does not loose information.

Apply this concept in media storage: An old cassette - tape has noisy and distorted sound, if you copy a tape and from the copy you make another copy, the sound will degrade a lot. A CD or an mp3 file on the other hand has information comprised of "highs" or "lows". Its copies do not degrade.

Signals treated as carrying "high" or "low" states are called **digital**.

DIGITAL LOGIC

The idea of being either "high" or "low" has the magic any two things have: If it not the one it is the other. That is what makes our computers! There is a whole logic or algebra (simple enough for an

8yo child) that can be made with two states, called Boolean logic from Mr. George Bool, a mathematician in 1850's.

Let's give to "low" the number 0 and to "high" the number 1. From now on we have 0's and 1's.

THE BINARY SYSTEM

10 states	6 states
0	0
1	1
2	2
3	3
4	4
5	5
6	10
7	11
8	12
9	13
10	14
11	15
12	100
13	101

Decimal	Binary
0	0
1	1
2	10
3	11
4	100
5	101
6	110
7	111
8	1000
9	1001
10	1010
11	1011
12	1100
13	1101

Humans used their ten figures to associate counting of physical objects. If nature made us with 3 figures in each hand we would count to 6. For seven or more we would add one more "digit", so we should count like on the left table. Technically they are called "Base-10" and "Base-6" numbering systems. The base-10 system we are used to is also called decimal.

How about having only 2 states? ("Base-2"). That should allow only 0s and 1s. That's the binary system. Let's count in Binary (on the left).

There is a saying: "…there are 10 kinds of people, those who understand binary and those who don't".

Binary system uses the magic of "if it is not 1 we know what it is, it is 0" and vice versa. It also "glues" with electronics such as switches states: ON or OFF. As digital signals that they are, we associate them with a voltage range, e.g. anything between 0 to 1.4V is "0", anything between 1.8V to 3.3V is "1" (not guaranteed to 0 or 1 if in-between) or analogously with 5V (3.3V or 5V is the supply voltage of our circuits). Each digit carrying a binary value is called **bit**. A group of 8 bits representing together a number (8 bits count up to 255 or have 256 combinations including zero) is called a **byte** (a number of n bits makes 2^n combinations like a decimal number of n digits makes 10^n combinations).

1.8 One plus one makes 10: the binary system and the digital world

Using binary numbers, or using decimal, or any other number base has no difference in performing the usual mathematic operations like addition, subtraction, multiplication, division etc. in binary 11+10 makes 101 just us 3+2 = 5. It is just another representation for the same quantity. But since bits can be literally the states of MOSFETs, real circuitry can now handle numbers (binary) and if complex enough, do such operations. Moreover its calculations can be ideally accurate if no 0 becomes accidentally 1 or vice versa (digital electronics are reliable enough to offer this), something that would be impossible if analog signals were handled.

We will not talk about logical gates yet as all classical books do. In nowadays circuits, you will find one gate every 20 circuits or more… technology goes on with microcontrollers. But we will see them in chapter 2.10 for theoretical knowledge with applications in software.

1.9 MICROCONTROLLER ANATOMY: EXPLORATION OF THE MAGIC CASTLE

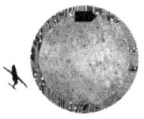

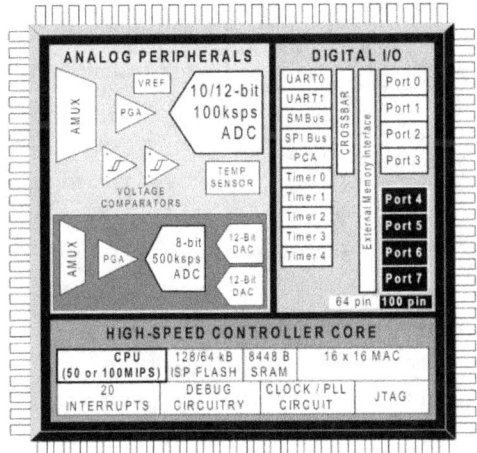

It's time to meet them! MCUs, (Micro Controller Units) the heart and the brain of most modern circuits. They contain a computer and lots of peripherals inside a single chip. They come in great variety of features. On the left we see only the basic of those.

The most complex components we encountered so far are the linear regulators, comprising of 10 to 50 transistors. Microcontrollers (MCUs) comprise of more than half a million. Their exploration will take two chapters for the most common features only.

A good thing is that 98% of their specifications and functionality description involves digital logic only (unlike many electronics-nature specs we found on other components). There is a common set of knowledge around "standard" functionalities they provide, in order to know how to select, understand and use the peripherals each one includes.

We live in a magic age where such ICs cost from 0.5$ to 10$ for buying in retail a single one. From year to year new MCUs enter the market with more features and computing power for less price. It is good not to stick in one part number for many years.

Let's imagine a microcontroller is a castle with many rooms and let's start exploring!

THE CORE & MEMORY

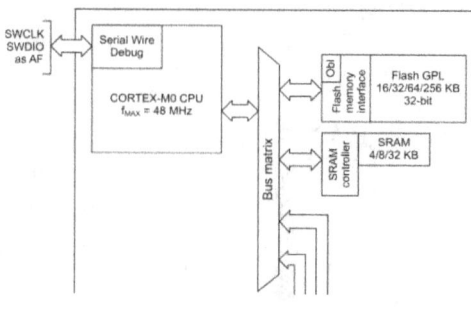

At the heart we find the computer. A CPU (Central Processing Unit) or "**core**" where the commands of our program are executed, memory of permanent storage type and volatile type. Permanent storage type memory (whatever written remains forever even when power turns off) is usually "Flash" type memory. Volatile type is a feature that we do not want (forgets its contents when power turns off) but unfortunately that is how RAM is (Random Access Memory), but we can change infinite times its contents very fast. In RAM we store changing data, in Flash we store the program and any permanent data contents.

How powerful is that computer? Well in absolute terms, for the jobs used, it is very powerful. They usually execute more than 10 million commands per second. Compared to our PCs, MACs or smartphones their performance is a joke, it is about 1/1000, but their cost (average 3$), size (average 5cm x 5cm) and the current they consume (around 20mA) makes them a treasure. Their memory sizes are from 8**Kbytes** to 500**Kbytes** in Flash memory (compare this to Terabytes of hard disks space) and from 256 **bytes** to 200**Kbytes** in RAM memory (compared to 2-8GBytes of year 2020 PCs and smartphones), but again their applications do not include playing YouTube in 4K resolution. They can store programs of many thousand lines of code and perform very complex tasks. Moreover they interact with the physical world by measuring voltages, generating signals, controlling physical objects like motors, lights etc. that PCs, MACs and smartphones cannot do, plus they fit inside a circuit.

The CPU and the memories had been the center of our castle. There are many rooms around, each is a peripheral functioning its own, doing very specific functions (if we use it). Each is connected directly to the core, sharing data or receiving commands from it (our program's commands)

GPIOs

The most useful and easy to understand peripheral is an internal controller that can use almost any of the chip's **pins** as inputs or outputs of digital signals. The pins functioning that way are called General Purpose Input/Outputs. Let's see this feature that is the most useful.

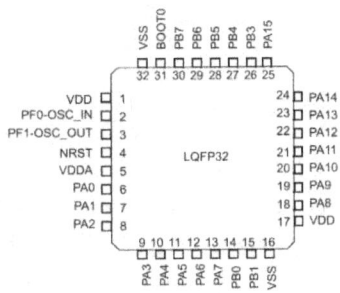

Let's consider a small MCU like the one on the left. It is from ST Microelectronics. ST named any GPIO pin as PXX (P from "port"). We can set any of those in the mode we want (input or output) initially. We can **read** the real time value of each **input** pin, which will be 1 if the voltage on it is more than the half of the supply voltage or 0 if it is less. The MCU supply voltage is usually around 3V (3.3V most times) but some older MCUs like we will see in the Arduino later, are supplied with 5V. If a pin is initially set as **output** (there are some output modes but we will skip them in this chapter) it will have a voltage on it that is 0V if it is **set** to zero or equal to the supply voltage if we set it to 1. Connecting an LED with its series resistor to an output GPIO is a first thought that works well. Each output can provide up to about 20mA to whatever is connected to it (*driven by it*). If our *load* is more current hungry we can use MOSFETs or other means to drive it.

ADCs

ADCs are a bridge between two worlds, the analog and the digital. The name means Analog to Digital Converter. Some pins can be **configured** to be ADC inputs instead of their default function to be GPIOs. Those pins can **measure** voltages (regarded as analog signals). Measuring means to convert a physical voltage into a number that our program can use. Their main specs are the resolution (usually 4096 steps), the voltage range (usually 0V to supply voltage) and the speed of measuring (converting to digital). For the latest they can perform around a million measurements per second. That way our computer can measure in real time physical quantities like sensors' output signals that correspond to temperature, pressure etc. Having a program of our own reading the states of GPIO input pins and ADC input pins in less than 1 microsecond and controlling GPIO outputs accordingly gives a lot of capabilities to create and maybe invent. Seat back and start imagining applications or examples you could make.

PWM

A pin can be configured in PWM mode that is Pulse Width Modulation, instead of its default GPIO mode. We will explain more what this is in chapter 2.8. It is an **output** mode that switches on and off fast in a way as to keep an **average** voltage on it anywhere from 0V to the supply voltage. That way we can adjust **how much** an LED will shine, how fast a motor will spin etc. The maximum current a PWM output pin can provide, again is a limit to what we can connect on our MCU GPIO pins directly.

UART

With the creepy name "Universal Asynchronous Receiver–Transmitter", UART or USART (S for "synchronous") is the de facto, easiest and most traditional way for MCUs to exchange data or **communicate**. Forget Gigabit networks or nowadays internet connection speeds. For simple communication where 0.1Mbit/second (equal to about 0.01MegaBytes/second) is enough speed, UART provides great two way communication using only two wires. Information (bytes) is transmitted through the transmitting wire (TX), bit by bit (8 bits in series, the one after the other) and can be received in the same manner on the receiving (RX) wire. Its history goes back to the 1960's and it is still the simplest way to communicate using a **serial** way of exchanging data. Unfortunately its old nature requires to set the speed for both the connected parties at the same rate to make it work. Devices to provide a USB to UART function for the PC are very common and

cheap (USB to UART dongles), so usually our PC communicates with our MCU over a UART. We can use it for the MCU to send any data, like measurement values, text messages and whatever else to our PC and view them on a console terminal, or for the PC to program our MCU or send to it data commanding it to do something. MCUs may also "talk" to each other that way. Note that throughout the rest of the book *"PC"* will be named any host computer, either PC, or MAC, smartphone or tablet.

Let's see quickly the rest of the serial communication methods. The main concept is that we spare 8 pins to transmit the 8 bits of a byte by serializing its (8) bits using as few pins as possible.

I²C

Another serial interface, a lot more sophisticated, thus complex, but with many useful features. It means Inter - Integrated Circuits communication, pronounced "I squared C" or "I 2 C" invented in 1982 by Philips Semiconductor (now NXP Semiconductors). It uses two wires. Its main feature is that it can connect many I²C peripherals simultaneously with an MCU. The MCU has the credentials of the **"master"**, connected peripherals are **"slaves"**. Each slave has its own unique **address**. The master asks for communication with a specified address slave, so it can select who it is talking to. Complex protocol but easy to use with supporting software. Its speed is up to 0.4Mbits/sec. Many sensors are providing I²C interface for communication with the MCU.

SPI

Another "Serial Peripheral Interface" that is simple, fast (up to over 20Mbits/sec) and needs 4 wires.

Timers

They count time, but they count really fast (they tick in frequency up to 10s Mega *Hertz* (MHz) (times per second), provide fast and accurate timing to our software or to pins that produce timing accurate signals or measure fast counting signals. PWM is produced by timers also. A special timing sub-system is called RTC from Real Time Clock. It takes the mission to count time when everything else of the MCU may be turned off, using extremely low energy from a coin cell battery.

USB

Some modern MCUs (not of the cheapest range) provide USB connection. That way our circuit can be plunged to a PC, operate as a USB device and usually be supplied from it.

Debugging interface

Most MCUs provide special pins (most times only 2 pins), and special "USB debugger" devices which connect to our computer that connect to those pins. Imagine writing an MCU program using your computer. MCU is not that computer, so you have to transfer the program to the MCU whenever you are about to test it. Testing involves finding and curing any problems called in software terminology "bugs" (from a real story of a cockroach in the first electronic computer ever made). Those magic debuggers provide "in-line" or "in-circuit" debugging that is your computer (PC) can command the MCU to stop execution and read any internal live data which you see on your PC as if you are programming and testing a program running in your own PC. A simper way is to use a UART allowing only to transfer your program to the MCU, not providing the "in circuit" debugging functionality.

1.10 ARDUINOS AND THE ARDUINO UNO BOARD

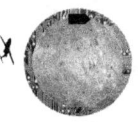

Let's leave for a sort while away from the technical stuff to relax and inspire with a few stories...

THE "TRIBE" OF SOFTWARE DEVELOPERS

Some people have been writing computer programs for a significant part of their lives. Let's talk about them, the software developers, or programmers or coders.

Software developing (or programming or coding) is a hobby of very high complexity. Envision a system of 200 <u>kinds</u> of commands (we include function calls here), two thirds already existing - which we need to know all about, one third of those kinds created by us (functions mainly) many data and 10,000 commands made by those 200 kinds. That's a medium sized program a skillful programmer can make alone, in a one to six months period. Do not freak out. In most practical electronic projects with an MCU, those numbers (for a medium size project) are around: 50 kinds, 500 commands. Moreover most projects are small.

It seems like all programmers, especially the self-taught ones, keep their minds in a good shape. Mind training and IQ raising by working effort is one thing. The other big thing is the nature of this "work". To people who love it (most programmers) it is playful, pleasant, making happiness exploding many times a day, inventive and a 100% creative. People who spend the most part of their lives in that mode (laughed by other people as "nerdy"), especially since their young age, they sometimes obtain some special virtues.

A percentage, perhaps higher than 5%, makes a personality that is altruistic. They have an idealistic view of the world, they want to help others and they want their life to leave a great footprint on this planet. It seems like the constant creativity (that is work, not birth talent) and all the mind pleasures of the process of coding spawns that. That process also never has tricky, nasty, sneaky, hostile people to deal with, but only logic and objectivity in every "cause and effect". Let us see some of those software developers: Elon Musk ("the goal is to make life multi-planetary"), the founders of Wikipedia (open knowledge), the founders of Twitter (open speech, open news), Richard Stallman (Free software movement), Julian Assange (corruption fighting), Linus Torvalds (Linux) among many others, just for making a picture to what we are talking about. Such people are on any field, but among software developers they are very common, mostly at making goals different than net profit. How many business do you know to have grown in the last century and back, which did not have the profit as their only goal? How many people have you heard of who prefer a good cause over a million dollars?

OPEN SOURCE SOFTWARE

This great "breed" of programmers tends to help fellow programmers or non-programmers who do not have access to expensive software tools which are useful to their (programming) lives. They share their creations (in their free time, it cannot be on paid time) with all the world. That software is the **open source software**, where the software author entitles everybody to do anything they want to with that software, as if it was theirs, in other words it is free. Most importantly they provide full access (there is nothing "locked") in order for others to continue its development or do whatever change they need and as much explanation of how it works (documentation) as they can. A notorious such software is the Linux operating system kernel that is over 25 million lines of code, all free to everyone.

THE ARDUINO IS ABOUT "OPEN SOURCE"

We have not said yet what is "Arduino". Some guys in Italy, around 2004, set in their minds to make open software that will tackle the following problem:

In the old times of "home computers" in the 80's, everyone programmed. That's because starting programming was easy and had almost no need to learn about other software code to "glue" their creation with it, in order to make anything working. As computers evolve, software tools inevitably have been turning more and more complex since functionality hugely increases, in the last 20 years they are over 100 times more complex than the "BASIC" language of those early home computers.

Those people (in the course of a University graduate program) set themselves to create a programming platform (programming tools such as programming language) what will be very simple, simple enough for non-programmers to be learned and used very quickly for creating simple software projects. The most important part was the "open source" philosophy since the beginning. They first made the project "Processing" that uses "C++" language with a very simplified extension in order to make simple and playful computer graphics. They took "Processing" afterwards and "ported" it into an MCU, making another very simplified extension to allow the control of the MCUs peripherals and CPU functionality. They called it "Wiring". That was accompanied with a really simple application (actually too simple) used by the users to write the MCU software (like the "notepad" we use in Microsoft Windows to write text), called IDE (Integrated Development Environment). All open source of course. A project spawned from the Wiring project, called Arduino named by the name of a bar

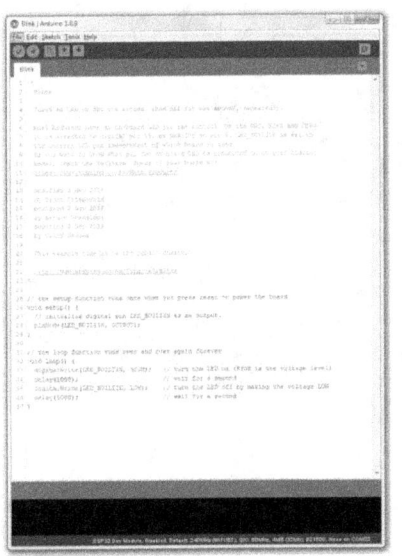

where some team members used to hang out. Thousands of developers liked the idea and the cheap hardware it used. They started using it and made their contributions on that open software, making it nowadays a very big open "library" of ready to use pieces of code for hundredths of functionalities of the MCU and of other electronics connected to it (e.g. screen displays, sensors, etc.). That's more or less the Arduino project. Its 95% about open source software, oriented for "programming the easy way". The other 5% is hardware, open "source" also (free to use designs or "open hardware"). The core idea of it is that someone who is not a programmer or a professional programmer can start-up coding and do useful things very easily.

ARDUINO BOARDS

Arduino compatible boards are electronic circuit boards with an MCU that is supported by the Arduino open source software. Historically they got into the market by the Arduino team who are still selling the original ones. Now they are sold from over 100 different manufacturers all over the world since anyone is free to make them and the design is available. There are many different models out there. The most standard one, since it was the oldest one, is the "**Arduino UNO**" model. Here it is:

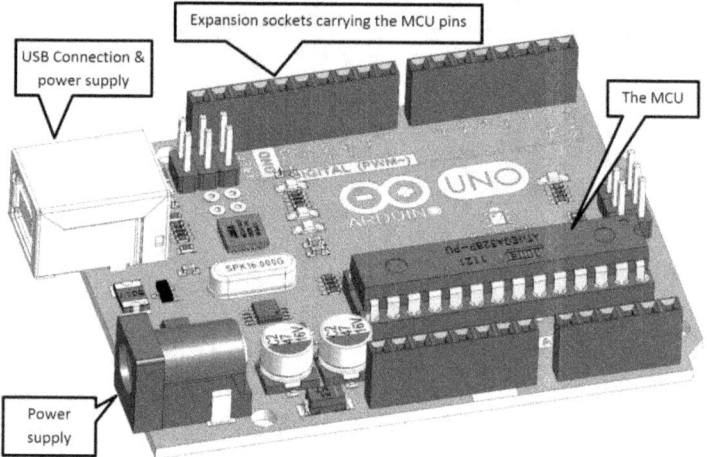

An Arduino board offers an easier way to use an MCU rather than having the MCU chip alone, namely:

- It is easier to connect components on the MCU pins through the expansion sockets called female headers (they contain spring conducts and are like the breadboard pin holes)
- Includes power supply circuitry consisted of regulators
- Includes about 8 more passive components required such as capacitors and a clock source of precise frequency (a crystal)

- Includes communications interface over USB (a UART to USB) to connect to a PC for downloading to the MCU our software and for exchanging data over a serial port.
- The Arduino Uno expansion pin headers position has been standardized, allowing others to make and sell expansion boards that fit on it, connecting with a single "snap". They  are called **"shields"** and provide many kinds of functionality, like the "Ethernet shield" on the left.

Besides Arduino UNO, there are many other boards, most with different MCUs which support the Arduino coding platform. The choice about which one to build with is about cost and capabilities. If for example our project needs more pins to drive or measure 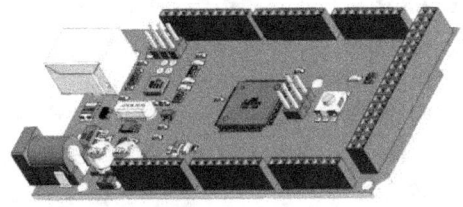 many resources (e.g. many sensors), a next choice might be Arduino **MEGA** 2560 seen on the left that is pin compatible to the UNO. If our  project is small and any dollar put to it is under some consideration or if the space is tight, we might go with a **"nano"** form factor as seen on the left. Besides the form factor, there is an MCU variety to choose from. The MCU choice has to do with performance characteristics, Arduino software compatibility (some may not be 100% but be 99% compatible to the Arduino software) and cost. Such decisions shall be taken after you finish reading this book. For now, let's introduce you to the **www.arduino.cc** web site and have a quick look around if you like. To those guys we owe a lot, so donating them the money for a beer sometime or purchasing anything from them may be a great reward they surely deserve.

ARDUINO UNO PERFORMANCE

Arduino UNO is unfortunately an old design, but still inside around the 50% of nowadays Arduino projects. Dated around 2005, its

MCU is not so powerful both in functionality of peripherals and in "computer" performance regarding execution speed and memory. Same applies for Arduino nano that has the same MCU. It is the part number ATmega328P from the former Atmel company, now Microchip Technology Inc. The board (UNO) costs around **5\$** though (from Chinese retailers or ebay.com), so it's not any great investment to make. Its worth because it is the most "compatible" or "standard" regarding expansion hardware (shields) and software. Currently it is under version (release) number 3, or R3 (previous releases are to the 98% the same). Let's take an idea of how much powerful it is.

- CPU: It has 8 bits bus (PCs are now 64 bits) with a clock running at 16MHz (compared to 3GHz of PCs), single core of course, able to execute about 16 million commands per second. For automation and measuring things it is much more than enough. For running a browser or playing multimedia it is out of any discussion. Since it is a computer that runs only the program we have written (there is not even an operating system in it), surprisingly the "lagging" in execution is a thousand times less than in a PC. There is no case of interruption between two commands execution for doing an antivirus scanning for example. It's a "hard real-time" system, where we have full control.

- Memory: **32Kbytes** for program or permanent data memory (Flash), 1Kbytes of a faster permanent memory for data (EEPROM) and **2Kbytes** of data memory (RAM). Here is mainly the problem of the "old horse" ATMEGA328 MCU it carries. It can store about 2000 lines of code or more that may be a reach program, but the data memory (RAM) is really small. It can barely hold the text (no photos, no formatting) of this page of the book you are now reading.

- Peripherals included as quoted in the datasheet of the MCU (ATmega328P): (you don't need to understand all of them at this moment)
    - *Two 8-bit Timer/Counters with Separate Prescaler and Compare Mode*

- One 16-bit Timer/Counter with Separate Prescaler, Compare Mode, and Capture Mode
- Real Time Counter with Separate Oscillator
- Six PWM Channels
- 8-channel 10-bit ADC in TQFP and QFN/MLF package - Temperature Measurement
- 6-channel 10-bit ADC in PDIP Package - Temperature Measurement
- Programmable Serial USART
- Master/Slave SPI Serial Interface
- Byte-oriented 2-wire Serial Interface (Philips I2C compatible)
- Programmable Watchdog Timer with separate On-chip Oscillator
- On-chip Analog Comparator
- Interrupt and Wake-up on Pin Change

In our days those are mediocratic features but they enable to build a lot of stuff. The flight computer on the space shuttle was not grater. Nor the computers they had onboard the great Apollo spacecrafts when they flew towards the moon. In chapter 2.9 we will see other Arduino compatible boards with MCUs that are a great deal more powerful. The "de facto" Arduino UNO's simplicity is a good start.

Another ghost of the past that it carries – which is a pain in design – is the voltage supply level. The Arduino team chose to use **5V**, as that was the standard of the digital electronics of that time, now the standard is **3.3V**, so we find problems when connecting some 3.3V hardware with maximum allowable voltage of 3.7V, to 5V signals.

CONNECTING COMPONENTS TO ARDUINOS

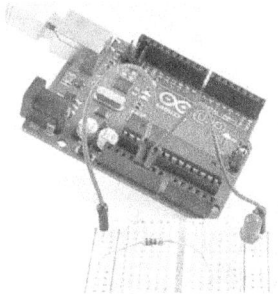

Finally, to grab even more the idea of how useful an Arduino board can be, let's see how we can practically connect an LED to it and control it by software (on the left). The photo is accredited to the greatest site for Arduinos, after the www.arduino.cc, the www.adafruit.com website that has flooded the Arduino open source library with its own contributions. (Credit: Simon Monk, https://learn.adafruit.com /assets/2158).

That job was done using the easiest and most popular method to connect anything to an Arduino board. **DuPont cables**! They are cables that come with a socket on each of their sides either female (hole) or male (pin).

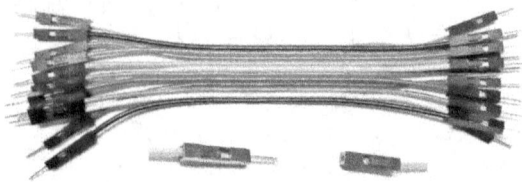

Those plugs directly (female to male) to an Arduino expansion socket (called header) or to a breadboard or to other boards with headers like sensors etc. we will see next. So there are four kinds of DuPont cables, male to male, male to female, female to male and female to female. They also come in various lengths. Surely your drawer of expendable stuff should always have some.

Regarding breadboards, Arduino "nano" boards are very breadboard friendly (left).

Besides all those "quick and dirty" methods just mentioned, in chapter 2.4 we will see how to make professional grade devices with Arduinos or make Arduinos ourselves.

1.11 SENSORS AND THINGS THAT MOVE STUFF AND DISPLAY STUFF

A brain (MCU) is useless if there are no eyes, ears and other "inputs" or "measuring instruments" as well as "actuators" like feet, hands etc. In this chapter we will have a glimpse of such input and output stuff as to start getting an idea of what useful creations we can make with electronics. The research and learning the technical details of each one are to be done from you, when you are about to use a specific one on a project. We are about to see a big world of capabilities all of which we can easily give to our circuits.

Let's start with the most playful, things that do actions, like motors, lights, displays, speakers etc. This chapter will be quite a big list of lots of toys. Take a break between them if you get dizzy here.

A RANDOM LIST OF SYSTEMS WE CAN CONTROL:

Motors:

The simplest motors we may know of, are the **DC motors** or brushed motors. Applying voltage on their poles (within the specified limits) the motor turns clockwise or anticlockwise according to the polarity of the voltage. They usually spin at about 2000 to 20000 RPMs (rotations per minute) which is usually too fast, so they are also sold with a

gearbox (a set of gears reducing the speed and increasing the torque or "rotating force") providing a more usable rotation speed, usually around 100 RPMs. Their mechanical power and thus the current they consume varies. Most usual ones, around 7cm long, with a gearbox, used for toys mainly, cost from around 1$.

They consume around 3A maximum (when powered up and kept stopped). How can we drive them (control them) with an MCU? We need a circuit that is also ready as an IC, called **H-bridge**. A good thing is that we can buy ready boards with such chips and all components included for costs starting at less than 1$. Their specs are the maximum current output and the maximum voltage input. Here is how they work: They "channel" the current of a powerful enough power supply or battery, to the motor's poles according to GPIO control signals of an MCU. Four switches are in the form of letter

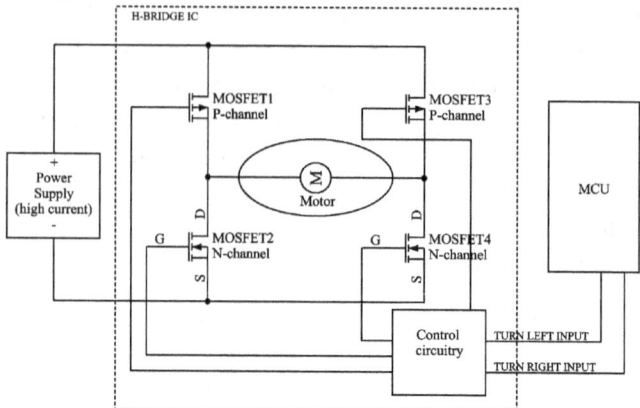

"H", in our example turning ON "1" and "4" will spin the motor to one direction, turning ON "2" and "3" will spin the motor to the other, non or "1" and "3" or "2" and "4" ON will stop it.

When we need motion, we usually need to move something to a new **position** rather just spinning it. **Servo motors** are the usual solution to this. They are mainly used in Remote Controlled (RC) models but later they have shifted their application to the robotics. Most of those can position their axis to a 180 degrees ranging angle that is set to the servo motor by three signals. + and − voltage supply (usually around 5V), capable to deliver at least 1A current and a control signal that uses timing to define the position (0° to 180° usually). That signal is created from a GPIO pin of an MCU. No extra circuitry is required since the servo motor contains the controlling circuitry. They stay on the position set, if the position changes they move to the new one in less than a second usually. Their specs are the force they can apply, their size and the mechanical quality (e.g. metallic or plastic gears). The cost for small ones starts from around 1$.

A last kind of motors used for positioning are the **stepper motors**. They offer very precise positioning and free spinning under controlled speed (they are not bound to 180°) only. They are called stepper because they are commanded to make one step at a time, clockwise or anticlockwise. The step size is around 2°. Making for example 1000 steps at one direction, if the step is 1.8° they will rotate by 5 circles precisely (1000*1.8/360). If they make again 1000 steps to the other direction they will return exactly to their initial position. That accuracy of motion makes them the motors of choice to move the printing head and the paper in desktop printers, 3D printers and to move accurately almost all industrial robotic

systems with ultimate precision. They need a "stepper motor driver" to be driven and sequences of a few MCU GPIOs to create the necessary control signals. Those driving boards are mainly specified by the maximum current and voltage they can provide. H-bridges also can do that job. The stepper motors are specified mostly by the holding force (torque) and the number of steps per cycle they provide. Their cost depends on their size, ranging from 0.5$ to 100's $.

Relays:

Imagine mechanical switches that do not change position by hand but rather change their position **mechanically** by an electric signal. Such devices (relays) can be used to control by "logic" of software in an MCU any electrical circuit connected to that switch. Switched circuits may be of any range of voltage or current

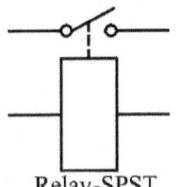

Relay-SPST

according to the switch's specs. We can turn ON or OFF devices plugged in the electric grid (220V AC or 110V AC), small ones and big ones (e.g. from a lighting bulb to electric ovens, heaters and pumps), also turn on/off devices working on low DC voltage consuming a lot of current like motors, heaters and any current hungry or high voltage device we may think off.

Relays invention is over 100 years old. They are mechanical switches with a lever that is not moved by hand, but it is internal and moved by attraction to an electromagnet (coil) whenever current flows through it. The switch conducts are electrically isolated from the coil's poles. Relays come with most kinds of two position switches (SPST, SPDT, 3P2T, 4P2T...). Their main specifications is the switch's conducts maximum current (usual is around 2A-10A) and the coil's voltage needed to activate (3V to 220V). They are mounted on circuit boards (PCBs) or on sockets (base) providing terminals to easily screw cables on them.

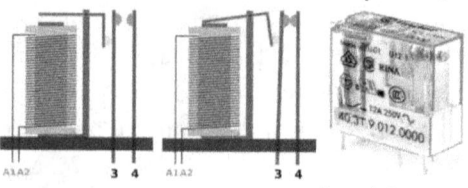

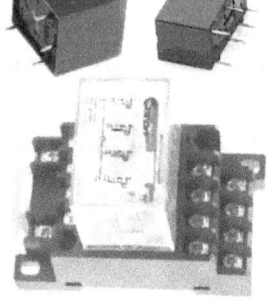

In order to activate their coil, the voltage applied to it (the coil) must be near a specified value, from 3V to 220V. Most usual coils are 5V and 12V. The current that has to be provided to it is usually around 100mA to 300mA. In order to provide such current, an MCU GPIO pin needs the help of a MOSFET or a BJT transistor. It also needs a diode in a tricky connection to protect our active component. It is handy and money saving to use relays with all this circuitry included, ready to accept GPIO signals costing around 0.5$ per relay, each providing an SPDT switch holding 10Amps. Another variant of relays are the solid state relays or SSRs which are fully electronic, handling only electrical grid AC Voltage. It is needless to say to take care to avoid electric shocks when connecting any wires connected to the electrical grid to a relay's conducts.

Displays:

The oldest and simplest type is the **7-segment** display. It is a digit comprised of 7 segments each of which is an LED (usually there is

an 8th segment that is the dot). In many cases if a GPIO pin is connected to each segment, it can provide enough current to illuminate it adequately, with a resistor in series of course. The 8 LEDs have internally connected all their Anodes together (common Anode) to a pin or their Cathodes (common Cathode). They come in various sizes and colors.

An evolution of the 7-segment is the **LED matrix**. It is an array of

LEDs (usually 8x8) configured in a kind of an array (16 pins drive 8x8 LEDs). They come in different number of LEDs, colors and size. We can place such components next to each other to make seamlessly a bigger matrix. Special controller chips in most cases are used in order to occupy less GPIOs from our MCU and to provide adequate total current to all those LEDs.

In chapter 1.3 we mentioned the RGB LEDs. RGB LEDs are a Red, a Green and a Blue LED encased together. Their Anodes are connected together to a pin in a "common Anode" LED, or their Cathodes in a "common Cathode". An evolution of those are the

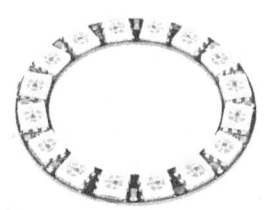

"**Neopixel**" LEDs. They are RGB LEDs encased together with a controlling IC with part number WS2812 or **WS28xx**. That IC accepts pulses of a single wire to set each of the 3 LEDs (R-G-B) to the required luminosity. If more pulses are provided, each such IC starts outputting the rest of the pulses to others connected like a chain "after" it, making it possible to control the exact luminosity and color of many such LEDs with a single wire (GPIO). They cost from 0.1$ to 0.5$ per LED. You may even find long LED strips or arrays containing up to 100 LEDs or more.

Getting away from the LEDs, proceeding to more advanced displays, we go to screens. The simplest are small **monochrome display modules** sized from 0.8" to about 5". They are called

modules when they contain a driving circuit and the display. They are divided in character and graphic modules. The first accept text information and display it with a fixed font only (left), the later control individually each of their pixels. Since most MCUs are low in memory and CPU power they are best fitted with a low pixel count display. The smaller are also the cheaper. The most widespread are the LCD (Liquid Crystal Display) and the OLED (Organic LED) technologies. The first use an LED backlight and work as "light valves" adjusting their opacity

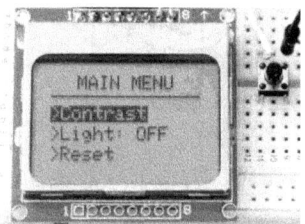

or how much of the back-light passes through. The later are self-lit LEDs, each pixel emits its own light. A commonly used LCD graphics module is a screen from an old Nokia phone, the 5110, shown on the left, taken from a video of the great YouTube channel "educ8s". It costs around 2$. Their size is usually from 1" to 4".

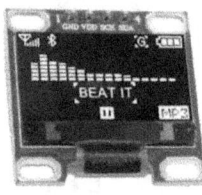

OLEDs on the other hand have vivid color (white, blue or green) giving a more impressive image. They are usually small, an average size is around 0.8" only costing around 2$. They range from 0.6" to 2". One less known issue they have is the life time of each of their pixels. If a pixel shines at its full intensity, after about one year of continuous operation it will reduce its brightness to about one half (counting lit time only).

Moving upper into the technology we meet **full color TFT** LCD screen modules. As LCDs, they use a backlight and act as "light valves" whereas now each pixel is comprised of three "sub-pixels", a red, a green and a blue. Each controls its intensity level to more than 100 steps rather than turning only ON or OFF. They range from about 1" to 7", usual sizes used are around 2" costing around 4$ (the

2" size). In sizes over 2" they are optionally sold with a touch panel that informs the MCU where we pressed it (pressure sensitive, not touch sensitive). (*Picture on the left is taken from www.adafruit.com*). Connecting such a screen to an Arduino will not enable you to do what a smartphone does. A single image requires usually more memory than an MCU has, so their applications are limited. The software needed for such functionality is big and complicated. Either they are limited to display simple images, or strong MCUs are used, or a TFT LCD with included "strong" MCU is used that provides ready functionalities to a "weaker" MCU. One of the latest kinds are the "Nextion" displays from Itead Intelligent Systems. Lately we see in the market high quality IPS TFTs which have a lot more vivid colors and 10 times greater viewing angle.

Sound:

Sound can be produced from simple "beeps" to playing mp3 tracks on a good quality loudspeaker. The simplest are the **active**

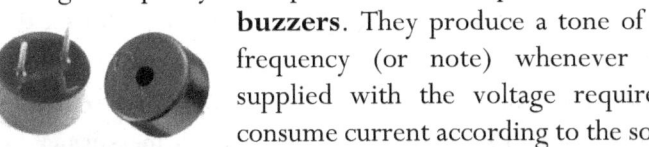

buzzers. They produce a tone of only one frequency (or note) whenever they are supplied with the voltage required. They consume current according to the sound level

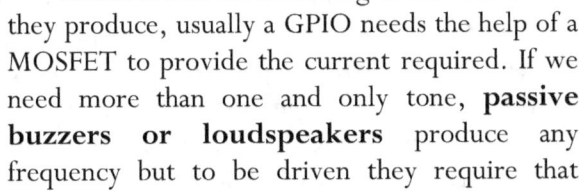

they produce, usually a GPIO needs the help of a MOSFET to provide the current required. If we need more than one and only tone, **passive buzzers or loudspeakers** produce any frequency but to be driven they require that frequency of sound to be fed to them as the same frequency of variating voltage. A MOSFET will do the driving job for simple tones sounds, an **audio amplifier** will be needed to generate any sound like music or voice. According to the sound level we require, the size and the price of a buzzer or speaker varies. A usual cost for hearing something within a silent room is less than a dollar. Finally, since MCUs are not fit in memory size and speed to playback sound,

 some boards that can play small tracks of **sound stored in mp3 format** are handy. They can handle around 10 tracks which are selected for playback using GPIOs. They cost around 0.5$! and require a small capacity SD card. Using them with an amplifier (if they do not contain it) and a speaker we can make applications like the repeated recorded sounds we hear in an elevator. A note is that any speaker has to be rated in more Watts than the amplifier used in order to be sure it will not be burned.

A RANDOM LIST OF SENSORS:

Previously we took a glimpse of things we can command to light up, move or "actuate". We can call them **outputs**. Those together with a **"brain"** are missing what can read/measure/sense information. Let's talk about those **inputs** that look like they are endless, like there is one for whatever we need to measure. Let's take a very sort glimpse of the most useful and easy to use. The device that converts a physical quantity (like temperature, light, magnetic field intensity etc.) into a signal is called **sensor**. There can be electronic devices or ICs with a sensor and a circuitry that makes the sensor signal easier to read (e.g. amplifying it, transforming it to a digital numerical value etc.) which we will still call sensors for making our life easier.

A primer about measurements first: Since we will be talking about measuring, precision comes to the surface. In measuring, besides "counting" (e.g. how many people passed) all other measurements suffer from tolerance or **errors**. In engineering, if we measure a voltage of 1V and our instrument reads 1.0001V the measuring error is equal to +0.0001V. Mistake is the human error, if e.g. we measure another voltage instead. So error in engineering is not something embarrassing but it is a performance spec. Another tricky concept is that of **accuracy** vs **precision**. Accuracy is how close the absolute truth is approached, never reached, checked using very expensive instruments and procedures. Our instrument can be precise and not accurate if it is not **calibrated** well. For example an expensive thermometer may read a temperature that is 20°C as 20.102°C to 20.108°C over hundredths of measurements over long time periods. It will be precise to about ±0.003°C but not accurate to 0.1C. **Resolution** of an instrument (e.g. reading 20.10302°C has resolution of 0.00001 °C) much higher than its precision is misleading and is common practice in cheap measuring products.

Temperature and environment conditions:

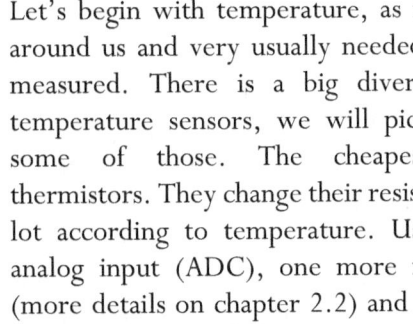

Let's begin with temperature, as it is all around us and very usually needed to be measured. There is a big diversity of temperature sensors, we will pick only some of those. The cheapest are thermistors. They change their resistance a lot according to temperature. Using an analog input (ADC), one more resistor (more details on chapter 2.2) and around 3 lines of code we can read the temperature with accuracy of around 1°C (1.8°F) to 0.2°C (3.6°F) depending on the thermistor and resistor quality. Resolution / precision is good (may reach up to 0.01°C/0.018°F easily). The measuring range is around -40°C (-40°F) to +100°C (+212°F). If we need to measure high temperature, thermocouples are the choice (left) together with an

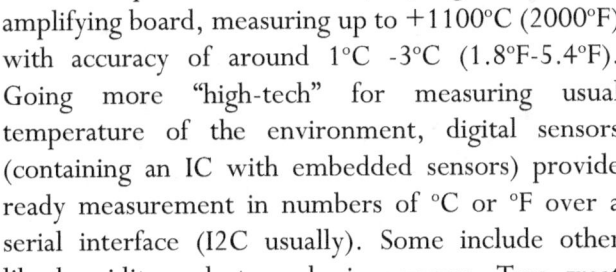

amplifying board, measuring up to +1100°C (2000°F) with accuracy of around 1°C -3°C (1.8°F-5.4°F). Going more "high-tech" for measuring usual temperature of the environment, digital sensors (containing an IC with embedded sensors) provide ready measurement in numbers of °C or °F over a serial interface (I2C usually). Some include other sensors too, like humidity and atmospheric pressure. Two great examples of those are the DHT11 and DHT22 (left) measuring temperature and humidity, costing 1-2$ and providing accuracy

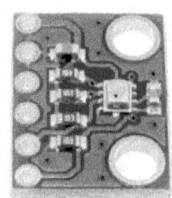 around 0.5°C (0.6°F) and the BMP280 (left) from Bosch. BMP280 measures temperature with 0.5°C (0.6°F) accuracy and atmospheric (barometric) pressure with stunning resolution (precision) of around 0.0001% or one part per million (1ppm) making it an altimeter of resolution that is better than 1 meter in altitude measurement. Small boards with it that connect easily to an MCU cost around 1-2$.

Light:

Measuring the environment light level has a few applications. As we will see later, measuring the invisible Infrared light (IR) has a lot more.

 The simplest and less expensive way (around 2c each) is the photoresistor (LDR), shown on the left. It varies its resistance according to how much light shines on its surface and – as the thermistor – we can use one more resistor and an ADC input to measure this. Its resistance is around 1Mohm in pitch dark and varies a lot in low light illumination: it drops to about 50K in dark indoors lighting conditions, goes to 1K in normal indoors conditions, to around 100Ohms in cloudy outdoors and to some Ohms in direct sunlight. Its precision is bad but it is handy to distinguish night from day for example or direct sunlight from shadow. If we need to measure the light (ambient light usually) more precisely there are photodiodes, phototransistors (we will not come to analysis about them) and IC light sensors which are the most  accurate ones. The most advanced provide accurate light and color measurement for applications to adjust smart lights illumination level and color (light temperature) according to the environment's light conditions, photography and others. Cost is less than 5$.

Distance of an object:

On the left you see the most recognizable sensor of DIY projects and educational robots. It is a distance sensor that uses ultrasound to measure distances from 3cm to around 3 meters with 1cm accuracy and a few millimeters resolution. Ultrasound (hence "ultrasonic distance sensor") is sound in frequency (pitch) higher than we humans can hear, it is usually around 33 to 40 KHz. Two signals are required, one output triggers the sensor to transmit a sort ultrasound burst (like a "ping" in submarines sonar, yes this is a sonar also) and an input that listens for an "echo" response from the sensor. Sound travels at around 340m/second. Counting the "trigger" to "echo" time provides the distance of a nearby object. This is called Time of Flight (TOF) technique. The sound "flies" the double of the distance since sound has to go from the sensor to the object and then back. That is around 59usecs per cm of object distance. The "viewing angle" is around 15 degrees. The cost is near 1$.

A latest technological advancing is light TOF distance sensors. They are IC digital sensors, providing distance numerically in millimeter accuracy. They measure the time of flight of light instead of sound. Light travels about a million times faster than sound, so they do an awesome job when they provide mm accuracy. Their size is much smaller, their range is around 1-2 meters which degrades on sunlight a lot. Their cost is around 4$. The name they come by is usually "optical range finder".

Proximity detection:

Rather than measuring the distance of an object, in many cases we want to know if an object is nearby or not. This true/false information is a lot less demanding in accuracy and is usually made by shining a light and measuring if there is reflection of it. It is practical to use an invisible light for that job, the **infrared light (IR)**. There are IR LEDs and light detectors called phototransistors for that. As we mentioned in the chapter 1.3 about LEDs, IR is light frequencies lower than the frequency of the red color (the lower edge of the visible spectrum) or in other words, of longer wavelength than red color (around 750nm). IR LEDs are usually in 940nm (nanometers) wavelength and less frequently at 840nm. Phototransistors come with filters allowing only 940nm or 840nm to pass through them, leaving a black color within the spectrum our eyes can see. Phototransistors are simply transistors (BJT) open to incoming light. They present the property of becoming more conductive (having less resistance) the more light (IR wavelength) shines on them. It is practical to use boards like the one on the left which integrate an amplifier and comparator with selectable "threshold" of reflected light intensity, thus selectivity of distance of object detection. They provide an output that is true or false (0 or 1) for a GPIO input pin of our MCU. They cost around 0.5$. An inherent pitfall of IR light reflection principle is that a black object will reflect less light than a white object so deeply black objects may be missed. Another application is to detect if the object is white or black. Their range is a few centimeters. A variation of those is the photo-interrupter or optocoupler (left) that detects if a non-transparent obstacle interrupts the LED light shining directly on the phototransistor. This is used a lot in counting things like gear's teeth passing through the sensor.

If we place a magnet in the object we want to detect, then we may use **magnet** proximity sensors or sensors of magnetic fields. The most widely used and simple enough are the **Reed switches** (left). They are small pieces of glass containing two conducts of magnetic material in close proximity. When magnetic field is applied (a magnet approaches), the conducts are magnetized and touch each other. We find them enclosed in many forms like the ones on the left named "magnetic door switches". Their practical applications seem endless, some examples are: detecting if a phone handle is in place, detecting an open door or window for security alarm system, detecting if a refrigerator door is closed and countless

Magnet

others. Another sensor for the same job is the **Hall Effect** magnetic sensor. They require a supply voltage and offer a digital output pin. Reed switches have a life of about one billion ON/OFF transitions but in counting rotating objects sometimes that is not enough. They also suffer from vibrations. Hall sensors have infinite ON/OFF transitions, they are vibration tolerant, are more reliable and fast. They also come in analog output, where the output voltage they produce is proportional to the magnetic field they sense. A final note for all the previous is that the magnet has to have specified direction since they sense magnetic field in one axis only (magnetic field is a vector field i.e. it points from North to South pole). Reed switches sense on the axis along their body, hall sensors on the axis perpendicular to their body. If the magnetic field is directed vertically to their sensing axis they will not "feel" it. Some hall sensors have polarity (need North Pole facing them for example, while South Pole will not do) Reed switches don't care.

On detecting **proximity of humans** (like automatic doors and lambs do) two main methods are used, both relying on the fact that humans are moving. One uses the human's body temperature that

is usually different from that of the environment. It works like the thermal cameras do, any object emits infrared light (IR) that is more intense the higher its temperature is. This IR light is called "far infrared" because it is far away in the spectrum of the "near infrared" the IR LEDs are (its wavelength is around 10000nm). Sensors called "PIR sensors" are one pixel thermal cameras and detect changes of the overall

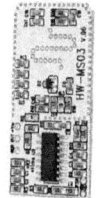

temperature which are caused by the (hotter) human's body motion. Those are prone to false detections by small animals or sudden changes in sunlight, the second method uses a real radar of microwave (>1GHz) electromagnetic waves but the radar does not rotate, it looks always at the same direction. Reflected radio waves of any moving object are magically making a "detection" signal (using Doppler Effect caused by the object's speed). Both systems cost usually less than 2$. They require only a digital input that is 0 for no detection, 1 for detection or even provide a relay switch conduct output. Note that it is controversial if the microwave radio waves of such low power cause any health trouble.

Rotary encoders:

Rotary encoders measure the rotation of objects like a shaft of a motor, a gear etc. If we place on a rotating shaft a transparent / opaque scheme like the one following and two photo-interrupters, as the shaft rotates we get the following signals (each is 1 or 0) with a respective "phase" or pattern labeling number.

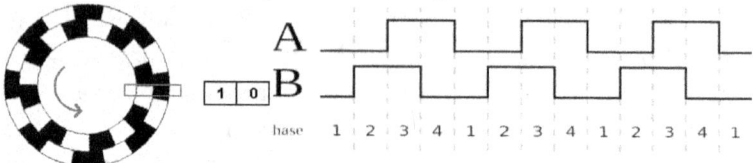

That way, if counting is increasing (1,2,3,4,1,2,3…) the above shaft rotates anticlockwise, if decreasing (3,2,1,4,3,2…) it goes clockwise. By dividing the circle in those 0/1 areas we know the angle we are at with precision (resolution) of 360°/patterns number. Usual encoders provide resolution of 100 "clicks" per rotation. The rotating pattern may also be "single" and use two adjacently placed photo-interrupters.

We may distinguish them in two categories. The first for measuring position precisely for robotics and stuff moving with motors. The second (left) for making knobs that rotate indefinitely to any direction for interacting with our hands with machines, such as selecting the volume to an audio amplifier. Those are also inside a computer mouse wheel. There is an even simpler form of encoders, those which measure "clicks per rotation" but not direction. Those have a single photo-interrupter or a single Hall Effect sensor and a rotating magnet.

Sensing direction and motion:

The following kinds of sensors are used for measuring what direction our device is looking towards: Accelerometers, Gyroscopes and Magnetometers. Each of those is a triple-sensor device, each is having 3 sensors vertical to each other. We live in a 3-dimensional world. Any 3 vertical to each other lines can serve as a "3 axis coordinate system" that can define any point in space with "coordinates" (named "x", "y" and "z"). We call such sensors "XYZ" or 3 axis sensors. So, following some basic physics, accelerometers measure the acceleration in each axis, gyroscopes (gyros) measure the angular acceleration (that is the "rotational acceleration) in each axis fast and fine enough and magnetometers surely measure the intensity of magnetic field (3 axis) acting as a compass usually. Our smartphone and our tablet find which direction the "down" is using mainly an accelerometer. The gravity of the earth we are living in causes a "downwards" like acceleration to all objects which we call "1 G acceleration". 1G is about (10 meters per second) speed change per second (10m/sec^2), it is present on any object on earth and is pointing downwards. Even if an object is still, the gravity force is like a downwards acceleration of 1G, so with a 3 axis accelerometer we usually sense the downwards direction (as a pointing vector) and

the tilt and roll "inclination" from the horizontal plane. A cheap and well established I^{2}C board with accelerometer, gyroscope and a lot of supported software is the MPU-6050 (left) costing less than 2$.

Detection of water and touch:

Water is not only the most common liquid, it is about the 70% of our body. How can you sense it? Many methods exist. Mechanical floats which activate a switch when lifted by buoyancy, resistance sensors that just measure resistance (air has practically infinite resistance as an insulator, water has some KOhms of resistance), capacity and a few others. About capacity, seeing it oversimplified, an open air capacitor (like two metal plates close to each other) increases it capacity a lot (about 100 times) if water is between its plates. Well water is the 70% of

our fingers so, taking this effect into much engineering, the "capacitive" touch screens of our smartphones are made. Touch sensors are single "touching detection" sensors that detect when the human body (our fingers usually) comes very close to them or touches them. Magically they are comprised of just a wire so they are maybe the only non-material sensors, but electronics are required in the MCU side.

Other:

Just naming: RFID readers (contactless card – tag readers), weight – force measuring sensors and precision ADC boards for such, sensors of various gases (CO_2, methane etc), pressure sensors for liquids or air, vibration sensors, sound detection, temperature measuring from distance using Far Infrared Radiation and more. Try some "Googling" in "Arduino sensor" or any specialty sensor if you are not covered up to this point. It is not enough to know if a sensor for a special application exists and how much it costs, you have to be able to use it. This book is about to provide to you the common and basic knowledge, enough to fill any gap yourself, unless your project is near the limits of the accuracy in technology for what you try to do and special scientific and engineering knowledge is required (e.g. try to measure weight in precision more than 1ppm, measure temperature with accuracy higher than 0.01°C etc.)

1.12 PROGRAMMING: THE BIG PICTURE AND ONE EASY PROGRAM

Let's relax our minds of this previous endless list of sensors, and go for a visit to a totally different world. Software. This chapter will assume you do not even know what programming is. We will only cover here an LED blinking program in C++ for Arduinos. If you can write this yourself, do a hop to the next chapter. If not, have fun meeting your first program. We will be down to practical reality, as we are used to. If you are about to practice this tutorial with real hardware, you need a PC (with Windows or Linux or Apple's OS X) and an Arduino UNO board with its USB cable. Nothing else.

THE WHOLE PROCEDURE:

Anything done for first time takes long time. All the procedure needed to be done in order to have our first testing program running should take you about 10 minutes if all go well (in order to install / setup things) and less than a minute after your system is setup to begin writing your second program.

Learning about programming Arduinos is a different story. You need to devote a few hours to understand the very basic stuff and there the road has just begun. If you are not a programmer you most probably need more than a year to become a well-educated expert in Arduino programming but the procedure will be 80% fun, 20% stretching your patience, 0% labor if you are challenged by electronics projects. This book is about to take you to the basic understanding and leave you in a level where you may continue that road yourself.

The procedure to be followed in order to be able to program is:

- Setup any software required (just one does the job)
- Connect our Arduino (we will use UNO here)
- Verify that our connection is OK
- Write – and test as you write – our own program

INSTALLATION OF REQUIRED TOOLS

As said, Arduino's mission is to involve a non-programmer or a beginner as easily as possible to creating things. Setting up what is required is about installing only one program which looks as simple as the Windows Notepad. Consider here that other programming missions (non-Arduino) require to setup a lot more stuff which are very complex and moreover to do 10x to 100x times more complex procedures in them in order to start writing your first line of code. The one and only application we need to get from the internet and install is the Arduino IDE. I, D, E stands for Integrated Development Environment and is the application that we use to write programs and do other programming relative operations.

The source for all (and most importantly the official reference) is the www.arduino.cc web site. In there, inside the "software downloads" section we find the Arduino IDE download page. It is assumed here that the reader of this book will know how to accomplish a "classic" application installation on her/his operating system (e.g. on Microsoft Windows, just follow the installer prompts pressing all the "Next" and "Yes" buttons). When this installation is finished, you should execute your new application and

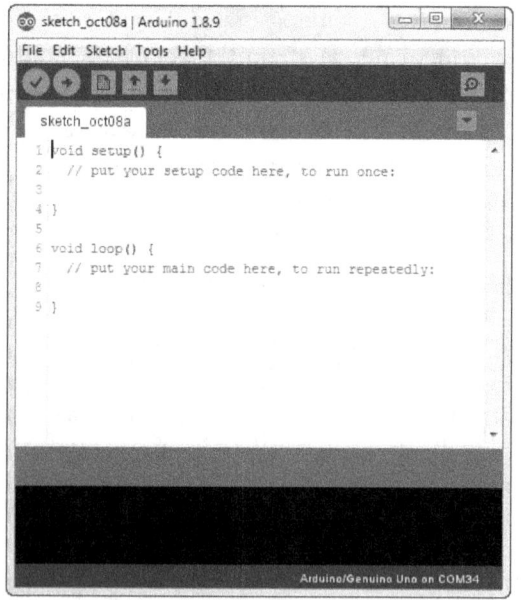

see it popping up (left). This is the – by design – simplest programming IDE. Hovering your mouse over the just 6 buttons you see on its toolbar, you see that they are: Verify, Upload, New, Open, Save and Serial monitor.

CONNECTING TO YOUR ARDUINO UNO

Connect your Arduino UNO to an available USB socket. Your device should be recognized by your operating system (you should not see an error anyway like "unknown device connected") after having installed the IDE that installs any driver necessary. Your Arduino UNO should also be powered up (some LEDs shining) by the 5V power supply provided to it by the USB bus.

By this moment a "serial port" device (that is a USB to UART device) should have been added to your computer. In Windows they are called **COM**xx, in Unix OSs like Linux or MAC OS X, they are called **dev/tty**.something. More on this next.

Open (if not already) the Arduino IDE, go to "Tools" menu and make sure the right board is selected

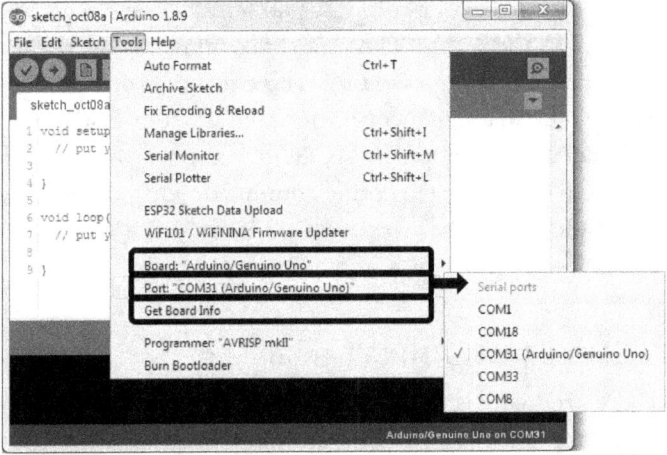

(Board: "Arduino/Genuino Uno"). Genuino is a name coined by the Arduino team later but everybody is still using their original one, Arduino. Then select the correct Port, which is the correct Serial Port. You should see a list of serial ports. If none is advising that an Arduino Uno is connected to it, you have to try each of them. After selecting a port, try "Get Board Info". If you have selected the right one and if all the driver installation works well (99% of cases) the message box on the left should appear. Otherwise an error message should appear. If none of the serial ports work (or none is present),

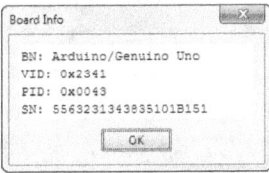

you should "Google" for your case (operating system etc.) and hopefully many nice guys will have posted a solution for you.

Having done that in success, nothing is between you and programming your board.

OK... WHAT IS "PROGRAMMING"?

Computers do some operations, unfortunately each of those operations is not complicated, but rather it is simple. Each of those operations is a **command**, stack many commands together and you have a program. Feed the computer (Arduino's ATMEGA328P MCU in our case) with the program, command it to start its execution (run) and you have an operating device acting as an alive computer. A program may finish doing what is intended to do quite quickly and leave our computer do nothing, or it may never end, as is usually the case. This set of commands is defined by the programing language we use. The execution of commands is done one by one (in a single core computer as in our case of Arduino Uno and the 95% of popular MCUs), when one finishes, the next is executed unless the executing command "jumps" to another one that is not the next. The execution flow can make various branches or loops.

WHAT IS A PROGRAMMING LANGUAGE?

There are many (more than 50) programming languages out there, about the 90% of all programming is done using the 5-8 most popular ones, one of which is the C++ used in Arduino programming. C++ bizarre name comes from its predecessor, C which is the most popular in MCUs programming. C++ is C plus some extensions (the ++ symbol increases the value of a variable by 1 in C).

Programming languages are not any close to human languages, just as computers operate so differently than humans do (yet). So it is unfortunate that a program could not be like the following:

1. *Connect to the Bank XXXXX computer*
2. *Hack it and access the clients' accounts data - Comment:(displaying the progress in the meanwhile like in action movies)*

3. Find the account number YYYYYYYY
4. Add to the deposit 1 billion USD
5. Disconnect
6. Say "program finished successfully. Quit your job and book a ticket to Bahamas"

But rather the available commands are quite stupid. It is the fact that many of those simple and stupid commands (thousands of them) do a little bit clever things. We use a programming language to write our programs. Any command or syntax that is not "known" to the language we are using blocks all execution with an error. So programming has strict rules of writing our program or "code".

What C++ commands may do

To take the first vague idea of the commands we can use in our programs, we will present some of the most used and useful kinds.

- Arithmetic operations (addition, subtraction, division, multiplication, other mathematical like square root etc.)
- Numbers comparisons
- Assign to "variables" numerical values or text
- IF something is true execute a part of code otherwise execute another (jump execution flow not to the next command but on another command elsewhere)
- Execute a part of code for some times (loops)
- Make our own "commands" called functions.

There are also Arduino specific "functions" coming ready as the Arduino "framework" or "library" which control any MCUs peripheral quite easily, like reading the input value or setting the output value of a GPIO pin, communicating data over serial interfaces (UART, I^{2}C, SPI etc.) counting time and do what almost all our MCU's internal peripherals can do.

Besides its commands, C++ offers ways to handle numeric data storage (in "variables" or "arrays") reusable code in functions and in classes, customized data structures and other many useful functionalities mainly for data storage and manipulation. That was a lot of theory... let's get to the real stuff.

SOME LANGUAGE SYNTAX FIRST

The "New" Arduino IDE window already contained the following program:

```
void setup() {
  // put your setup code here, to run once:

}

void loop() {
  // put your main code here, to run repeatedly:

}
```

The basics we have to know are:

- "//" denotes comment. Anything following this in the same line is ignored by the language but is useful for us to write notes. If we need multi-line commands we can use "/*" to begin and "*/" to end our comment session.
- `void setup()` is a **function**, beginning with opening brace "{" ending with closing brace "}". Same for the `loop()` function. The "()" parenthesis denote their parameters which in the previous are none. "`void`" denotes the kind of value they return, the previous returns nothing.

This program is a two empty functions program, therefore doing nothing. Let's fill them up with some commands. We will do it the easy way. Click the menu "file", submenu "examples" sub menu "01.Basics" and select Blink program from there

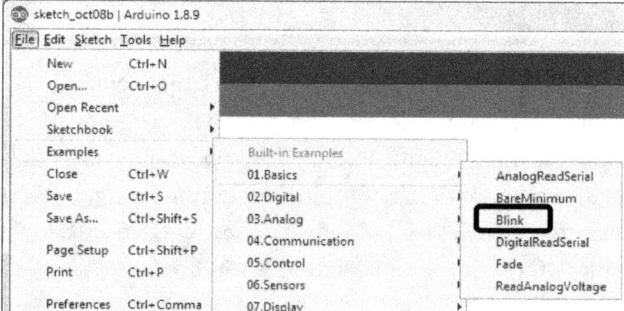

Ignoring the first lines of comments, this is the program that just appeared in our IDE:

```
25 // the setup function runs once when you press reset or power the board
26 void setup() {
27   // initialize digital pin LED_BUILTIN as an output.
28   pinMode(LED_BUILTIN, OUTPUT);
29 }
30
31 // the loop function runs over and over again forever
32 void loop() {
33   digitalWrite(LED_BUILTIN, HIGH);   // turn the LED on (HIGH is the voltage level)
34   delay(1000);                       // wait for a second
35   digitalWrite(LED_BUILTIN, LOW);    // turn the LED off by making the voltage LOW
36   delay(1000);                       // wait for a second
37 }
```

This is a complete program. Before explaining it, since practicality in this book surpasses theory, let's use it first and explain it afterwards. Just notice the same `setup()` and `loop()` functions right there again. Take one minute to read it and imagine what it will do.

Transferring our first program to the Arduino board and running it

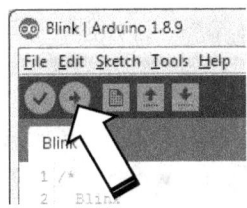

If your Arduino is already connected via USB, and the serial port is set, just press the "upload" button (left). If the serial communication is OK you will see a message in the IDE "Done uploading" and most amazingly you will see an LED on your board blinking once every second and forever. That's a joy of seeing a program running well on a new hardware…

Seeing that in practice, lets revisit this program to explain some functions used there we have not explained so far.

First, as the comments also point out (line 25 of the program) any commands put in the `setup()` run once when the system starts up. And as the comments point out in line 31, after the commands inside the `setup()` finish, execution goes to the first command (line) of the `loop()`. When execution point reaches the end of the `loop()` (closing brace in line 37) it goes again to its beginning (line 33) and that goes on forever.

What does `pinMode()` do? It sets a GPIO pin to input or output mode.

What does `digitalWrite()` do? It sets a GPIO pin already set in output mode to high or low. Low is 0Volts, high is Voltage equal to the power supply which is 5V

What does `delay()` do? It does nothing keeping the MCU waiting for some time to pass, counted in milliseconds.

Notice that every function or command we use ends with a semicolon ";". This is an annoying syntax of the language, we will get used to it.

If you have felt well and deeply how this works let's move on to play a little.

LET'S PLAY A LITTLE

Let's make the following: If a button is pressed the LED should be always ON, if not the LED should blink.

We have to connect a real button on a GPIO pin, lets choose randomly pin 12 (actually pin 13 was avoided since it is connected to the on-board LED) and either connect a button between this pin and its nearby Ground (GND) pin or sort those two together with a "U" shape piece of wire when we want to "play" with it. So, pin 12 will be either unconnected or connected to the Ground (0V).

Now let's set this pin as GPIO input (once, on program start in setup() of course)

```
void setup() {
  pinMode(LED_BUILTIN, OUTPUT);
  pinMode(12,INPUT_PULLUP);
}
```

Now let's read the GPIO #12's value in the `loop()` with the `digitalRead()` function and use the grail of the programming commands, "if" to do our logic. The whole program, comments removed, is here:

```
1  void setup() {
2    pinMode(LED_BUILTIN, OUTPUT);
3    pinMode(12, INPUT_PULLUP);
4  }
5
6  void loop() {
7    if (digitalRead(12) == LOW)
8    {
9        digitalWrite(LED_BUILTIN, HIGH);
10   }
11   else
12   {
13       digitalWrite(LED_BUILTIN, HIGH);
14       delay(1000);
15       digitalWrite(LED_BUILTIN, LOW);
16       delay(1000);
17   }
18 }
```

Explaining all: Pin 12 uses a pull-up resistor internal to the MCU. That is a resistor about 30KOhms connected to the positive 5V power supply. This holds its state (as input that it is) to 5V that is HIGH or "1" as long as our button is not pressed (pin 12 is connected to nothing).

Our button connects to the Ground (0 Volts), so as long as we press it, its voltage is 0V that is "LOW" or "0". Reading the lines 7 and after you should understand how "if" works. A final note is that: to compere if two numbers are equal we have to write

"`if (numberA == numberB)`", not "`if (numberA = numberB)`".

Of course, even with the little programming knowledge served up to this point, we can write our program doing the exactly same thing with at least one different way. Yes, many programs can do the same functionality. The best is the simplest and the more comprehensive by the author, by some others it is the most expandable to new functions.

1.13 WE JUST CIRCUMNAVIGATED THE ELECTRONICS PLANET, LETS LAND AND DO THAT AGAIN ON THE GROUND WITH A FAST CAR

So far we had our first taste of the electronics wonderland. At the next section we will revisit almost everything we said so far and more, looking at them from a closer point of view. Let's ride.

2. DRIVE FAST THROUGH THE ELECTRONICS WONDERLAND

2.1 VOLTAGE AND CURRENT REAL ENGINEERING

This world is not ideal, engineering is required almost everywhere. Mathematical theory applies without corrections almost nowhere. Starting our next tour with this terrifying quote, let's explore that in the most fundamental stuff of electronics, voltage and current.

Almost all signals are voltages. Voltage represents (practically in every analog signal) how much something is e.g. a physical quantity like temperature. If the voltage introduces errors (as we said in the "sensors" section of 1.11 about measurements) those errors add up to our measurements errors. Current on the other hand, either induces voltage errors or kills our circuits with a smoky death. Let's see the most usual traps we usually fall upon regarding bad voltage or current engineering and how to avoid them.

PITFALL #1: UNCONNECTED INPUTS

This pitfall has so big effect, that can make even a digital signal to deviate so much that may go from "1" to "0" or vice versa.

The case is to have as input, something that is either connected to nothing (is an open circuit), or is connected to something of defined voltage with a series resistor that is huge, tens of Mega Ohms or more. A classic such mistake is to read the value of an MCU digital **input** or the value of an MCU ADC analog input using the circuit of the left. When the button is pressed, the input is connected to a well-defined voltage (3V). But while it is released the input is connected to nothing. We call this state **"floating"** or **"tri-state"**. What defines the voltage of a wire (a circuit node) that "hangs in the air"? Actually randomness. In reality, on this example it may fluctuate from 0V to the supply voltage of the MCU (5V on Arduinos) due to internal protection diodes limiting it to this range. Physically it behaves as a receiver's antenna. It may change either about a hundred times per second at the frequency of the electrical grid (50Hz or 60Hz) by interference of nearby electrical cables, or change every some seconds to some hours by electric ions in the atmosphere or parasitic resistance of some Giga Ohms to nearby wires, caused e.g. by humidity, slowly charging the parasitic capacitance of the input (engineering is required almost everywhere...)

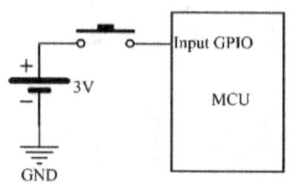

The usual cure to such situations is to use a resistor to "tight" our floating node to the ground (0V) or to the supply voltage. The circuit on the left solves this problem with a resistor big enough to avoid high current consumption while the button is pressed. Using 10K resistor for example, causes I = V/R = 3V/10000Ohm = 0.3mA current consumption while the button is pressed and also keeps the input voltage well at 0V while button is up. A variation is the circuit on the left.

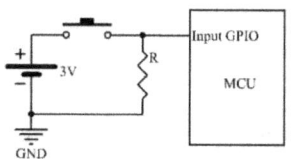

Supply voltage is usually called **VCC**. (5V in Arduinos, 3.3V in most MCUs). The input's value in this circuit is reversed, 1 while not pressed, 0 while pressed, nothing we cannot handle in software. Since this is a frequent problem, MCUs provide this resistor internally which is

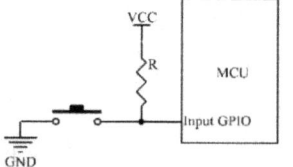

we can choose to be connected or not by software. Giving more terminology as we go, the resistor on the previous circuit is called **"pull down"** resistor and the resistor on the last circuit is called **"pull up"**. "Up" or "high-side" is the VCC, "down" or "low side" is the ground. We use to keep that in drawing also as much as we can, (left) placing VCC upwards and ground downwards. From

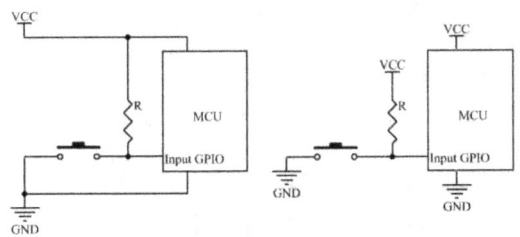

now on, forget batteries symbols, we will go with the VCC symbol in our schematics like in the last two circuits. The two last circuits are equal, VCC is called "net name", since it is in many places we spare to draw many lines of wire, all same net names are one and the same node.

Pitfall #2: Sort circuits and issues on voltage sources connections

Let's briefly talk about real (not ideal) **voltage sources**. Let's take a voltage source we all have some feeling of, a battery, say an alkaline AA battery. Let's also take a wire that is 0.001Ohms that is about a 2mm wide, 10cm long wire of copper. Sorting our battery with it makes 1.5V on a 1mOhm resistor. So we will see current of 1500 Amperes (with quite devastating sparks and magnetic fields)? Nop! Current will be about 10 - 20Amps with a fresh battery. This

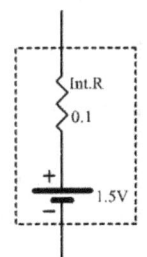

is why: Those batteries have inside them electrodes and chemical stuff which make a total **internal resistance** (that is parasitic / unwanted resistance) of about 0.1 Ohms when the battery is fresh, increasing to about 0.3 ohms at 50% of its energy usage. So, this battery is an ideal voltage source with a series resistor of a fraction of an Ohm. Coin cell batteries have about 20-50 Ohms, the lead-acid battery of a car has around 50mOhms.

What about **power supplies**? Those also have internal resistance (of less than one Ohm) but mostly important they can provide current up to a limit (specified on them). Exceeding that limit either their voltage drops as to protect themselves from burning, or they burn by overheating. That current specification is the upper limit of capability to deliver current. We should always chose a supply with 150% at least of the maximum current our circuit may ever consume. Last but not least, there is another voltage source that is on dangerous voltage levels (to our body) and has a lot of current to deliver if sorted. That is the mains power in our house. Sort circuiting it is not fan at all, but a nutsy and dangerous explosion. The bottom line is that we must always try as to make sure that a voltage source may never get sort circuited. **Fuses**, little devices that blow themselves when current flow exceeds a specified value help to achieve this, others need replacement for our circuit to work again, others are "resettable" and restart working when current is low again.

What about connecting together two or more **GPIOs** of an MCU configured as outputs? The MCU internal circuitry connects them to the VCC when set to "1" and to Ground when set to "0". Is this connection ideal, with zero resistance? Let's imagine that it is. On the circuit on the left in that case we have quite a big problem. If output A is 0 and output B is 1 what

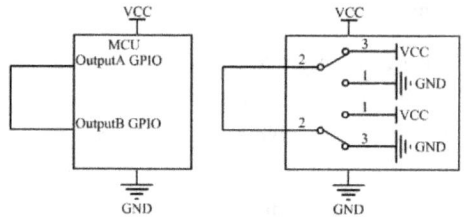

happens is that the power supply strangles at its limit. Moreover, if a high current capable supply is used, let's say, a big 10A supply, will the internal switches (MOSFETS) of the MCU handle such current or will they make a puff of smoke and decease? Their absolute maximum current specification unfortunately is below 100mAmps! (Each MCU has its own specs). In reality GPIOs MOSFETS have a resistance that is about 30 Ohms (differs from MCU to MCU) so that at least limits the current in such a situation, but yet to about 100mA Doing the previous example or that on the left is

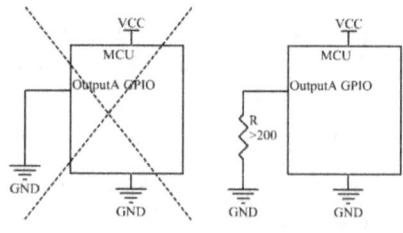

still a bad idea. It may burn all the MCU or that GPIO pin only. The safe current limit in most MCUs GPIOs is around 20mA. In such cases a resistor should be placed in series to limit the maximum current.

Pitfall #3: Wires too thin

The biggest consideration we take when the word "current" comes to our mind is: "is our wire thick enough for it?" Wires are almost always made of copper. Copper is of the most conductive metals but still it has some resistance. 1 meter of copper wire of 1mm diameter has resistance of about 20mOhms. Halving the length halves the resistance, halving its cross section area doubles its resistance. This unwanted resistance makes problems. High current flow creates an even higher voltage across our wire equal to $V = I*R$ (a "voltage drop" that we will visit in the next chapter). Another issue it has is that resistors (aka wires) heat up according to how much current flows through them (a topic we will visit this in the next chapter). If we use a thin wire to pass current through it, over a current value it will heat noticeably and over another, even higher value, it will melt and stop working (sometimes by fire on its plastic insulation if not a good quality one). The following table shows a picture of approximate current limits as well as the wire's resistance.

Gauge (AWG)	Conductor Diameter Inches	Conductor Diameter mm	Conductor cross section in mm^2	milli Ohms per ft.	milli Ohms per m	Maximum amps for chassis wiring
1	0.289	7.35	42.4	0.1239	0.40639	211
4	0.204	5.19	21.1	0.2485	0.81508	135
10	0.102	2.59	5.26	0.9989	3.27639	55
14	0.064	1.63	2.08	2.525	8.282	32
18	0.040	1.02	0.823	6.385	20.9428	16
22	0.025	0.65	0.327	16.14	52.9392	7
24	0.020	0.51	0.205	25.67	84.1976	3.5
26	0.016	0.40	0.128	40.81	133.857	2.2
29	0.011	0.29	0.0647	81.83	268.402	1.2
32	0.008	0.20	0.0324	164.1	538.248	0.53
35	0.006	0.14	0.0159	329	1079.12	0.27
38	0.004	0.10	0.00811	659.6	2163	0.13

In practice when connecting GPIOs of our MCU we shouldn't care of whatever our wire is, connecting motors or other current hungry devices should make us concern. It is very hard to see any fire below 2A current, a very sort puff of smoke probably should be all the fun. Be very careful of DuPont wires! Cheap ones may have around 0.2Ohms per 20cm length and hold maximum 2A.

PITFALL #4: EXCESSIVE ELECTRIC NOISE

Noise is random fluctuations of voltage. In real world it is everywhere, literally. Noise is the biggest pain in analog signals and analog electronics in general. It is created in amplifiers, resistors, ADC measuring inputs, and in sensors themselves, adding up to the measured signals. Noise defines the resolution capability of measurement. In sound systems it is an annoying "hiss" sound. Besides noise a signal can get similar unwanted additions by **interference**. We have interference when a wire acts as a receiving antenna to nearby electromagnetic signals travelling over the air. You may already have heard a humming sound in nearby speakers when a cellphone is ringing. Interference may also be present by mains power 50-60Hz alternating current or by nearby magnetic fields produced by coils, motors etc.

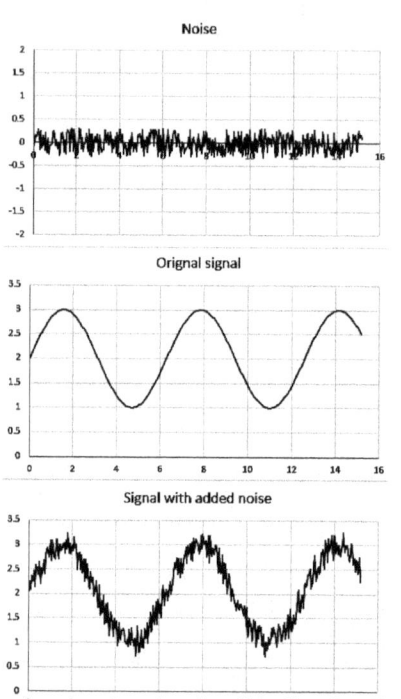

Noise is a big chapter itself. There are other pitfalls besides those said, but it is a big analysis to go all over them as they play less part in the game of deceiving us in the accuracy of measuring.

You may ask: Hey, you haven't said a word yet about AC/DC!! Patience...

2.2 RESISTORS RECIPES

About 80% of electronics design is about resistors and unwanted resistances. So far we have introduced Mr. Ohm's equation (I = V/R and its variations R = V/I and V = R•I). We will introduce only 3 more equations here which are really useful to all electronic designs. Mathematics as we said are kept as far away as possible but some very simple are in our way all the time.

VOLTAGE DIVIDER

Voltage divider is maybe the most important structure of electronics dealing with non-digital signals (call them analog electronics). It can't be simpler. Here it is, it's just two resistors in series:

And it's equation we have to know is:

$$V_{out} = V_{in} \frac{R_2}{R_1 + R_2}$$

(Middle voltage equals the input voltage by the ratio of the bottom resistor to the resistor sum)

They connect their "outer" ends to a voltage, their middle point is a fraction of the "outer" or supplying voltage. This is going to be a guy we will usually hang out with in electronics designs, so let's meet him better.

Let's place any two same resistors for R_1 and R_2. The output voltage will be $V_{in}/2$. If R_1 is zero, $V_{out} = V_{in}$. If R_2 is zero, $V_{out} = 0V$. To feel this circuit's operation better we may visualize that V_{out} (the "middle" point) "connects more" to the voltage V_{in} the less the R_1 resistor is and the more the R_2 resistor is and likewise it "connects more" to the 0V Ground point the less the resistor R_2 is and the more the resistor R_1 is as we may see in the table that follows. There

2.2 Resistors recipes

Vin	R1 (KΩ)	R2 (KΩ)	Vout
1	1	0(sorted)	0.000
1	1	0.1	0.091
1	1	0.5	0.333
1	1	1	0.500
1	1	2	0.667
1	1	5	0.833
1	1	50	0.980
1	1	∞(open)	1.000

we have as an example R_1 equal to 1K. Doubling both the resistors provides the same V_{out}. It is the ratio of the bottom to total resistor that matters only. Using very low Ohm resistors to do the same job (e.g. 1Ohm and 1Ohm for $R_1=R_2$) introduces the problem of increasing the current flowing from both the resistors. Using very high values seems to spare current drawing but too high values may introduce problems, as we will see next of R_2's value dropping by parasitic resistances connected parallel to it. Usual values for most jobs are in the region of 1K to 50K for total resistance (R_1+R_2).

Let's see an everyday met voltage divider. When we use a voltage source with an internal resistance (all batteries for example) or a voltage supply connected by a cable of non-negligible resistance (e.g. thin and long), we have a circuit like the one on the left. Assume we connect a device that is supplied by 5V and needs (consumes) 200mA. Such a device is like a resistor equal to R = V/I = 5/0.2 = 25 Ohm. Let's supply this by the USB bus's 5V using a long low quality cable that has 2 Ohms resistance (numbers are realistic). This arrangement will be like the circuit on the left. Is this a voltage divider? Yes. How much is the supply voltage on the poles of our device? It is 5V*25Ω/(25Ω+2Ω) = 4.63V instead of 5V. Where have the 5V − 4.63V = 0.37V gone? Well, they are the Voltage across the R_1 resistor. Current I is equal to I = V/R = 5V/(total resistance) = 5V/(R_1+R_2) = 5V/27 Ω = 0.185A. The voltage across R_1 is V = R*I = 2 Ω *0.185A = 0.37V!!! This is called **voltage drop**. If we apply the Ohms low in R_2 we have V = R*I = 25 Ω*0.185A = 4.63V = V_{out}! Almost no equation is required to be remembered as you can see since we can easily calculate stuff from more basic laws, but the voltage divider is very frequently met and it is practical to memorize it.

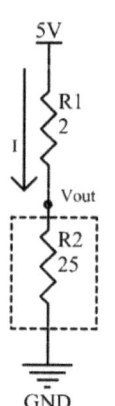

TOTAL RESISTANCE

As said in chapter 1.2, resistors connected in series behave as one resistor with value that is the sum of their resistance.

$$R_{total} = R_1 + R_2 + R_3$$

What about resistors connected in parallel? They behave as a resistor that is less than the smallest of those. That's since any parallel road the current can take, makes it flow easier in total, than having one road only. The equation is bigger here but do not worry, we will see very few more through the rest of the book.

$$R_{total} = \frac{R_1 \cdot R_2}{R_1 + R_2}$$

$$R_{total} = \frac{R_1 \cdot R_2 \cdot R_3}{R_1 + R_2 + R_3}$$

If $R1 = R2$, total resistance is the ½. If $R1 = R2 = R3$, total resistance is 1/3 etc. Another quantitative concept is that if we have a small resistor and connect parallel to it a big one (even if that is a parasitic resistance), the total resistance will lower of course, but will drop very little (e.g. 1Ω // 1KΩ makes Rtotal = 0.999 Ω - 0.1% drop). The opposite happens if our resistor is big (e.g. 10KΩ // 1KΩ makes Rtotal = 0.909KΩ - 909% drop).

HEAT AND POWER

There is a physical phenomenon that is useful in life and mostly unwanted in electronics. Current passing through any resistor produces thermal energy – heat. Old lamp's filaments are resistors heating up, so much that they reach a temperature of about 3000°C (5000°F). Electric heaters, cookers etc. use also resistors that heat up. Introducing some simple and necessary physics, resistors actually produce a flow of thermal energy, or equally, have a rate of heat produced per second. The produced energy per time is called **thermal power**, in our case it is calories or Joules per second or

Watts (after Mr James Watt at around 1760's, well, imagine the conversation: "- What is your name? - My name is James Watt. - What?"). Heat (thermal energy), heat rate (thermal power) and temperature (actually temperature rise) are different. Applying say a 50Watts (W) thermal power to a lamp's filament makes heat production rate highly "concentrated" and thus gives a very high temperature rise of around 3000°C. The same 50W in an electric soldering iron tip will make about 400°C and the same 50W on an electric heat radiator will make 1-5°C temperature rise.

Any resistor of same value (Ohms), no matter how it's made or its quality (a rusty wire, a transistor's resistance, the most expensive resistor in the world), will produce exactly the same heat rate or thermal power if the same voltage is applied on it (also the same current flows through it by the Ohm's law correlation of R, V, I). How much thermal power?

Power (Watts) = V•I

This is actually a conversion of energy from electric kind to thermal kind. In resistors (in any electric resistance) the 100% of **electric power** (energy per time) is converted into thermal power that dissipates in the environment. So in resistors electric power = thermal power. An opposite case is power supplies which produce electric power (also rated in Watts). Avoiding more equations, you can apply Ohms law inside the power equation yourself and replace either the V to I*R, or the I to V/R to get more power dissipation equations.

Why all those said? Resistors heat up, we do not want that, we cannot avoid that ever and we have to choose the proper resistor to be capable to dissipate that power without burning up. Resistors have a power rating that is the maximum power they can handle. The bigger size they are the more that is. We are good as long as we do not exceed the maximum power rating ever. A 10W resistor for example, will do the same function if any amount of power less than its rating is applied, e.g. if V*I = 0.01W. More on resistor sizes on the next chapter.

2.3 COMPONENTS: TECHNOLOGIES, SIZES AND WHERE TO FIND THEM

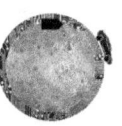

Having enough for theory about current and voltages, let's move to stuff we can touch and assemble. Components. Electronics are well inside the engineering regime. Knowing the Ohm's law but not knowing how real resistors are and how to get and use them is almost pointless. What someone can achieve in practical electronics design and making, has mostly to do with how much his knowledge extents to "what component is out there for the job and which one is the best" than to theory, if of course one has comprehended the theory basics. In this chapter we will take glances to most common and useful components.

We will see the world of components as a line with two ends and all the points in between. On the one end imagine components big sized, old school, very handy for making prototypes with breadboards, no matter how big or costly they are. On the other end imagine what is used by the biggest tech companies inside the latest smartphones or smartwatches. Miniature, most times requiring great optical magnification to work with them, complex in functionality and optimized for lowering cost. The road of electronics starts at the first end of this line and goes towards the other end as we may end up sometime designing real products that will be intended to go in the market, mass produced. The presentation following shows the most of this spectrum.

GENERAL CATEGORIES

Almost all have to do with the shape or the "package" used to host the functionality such as a resistor, capacitor, chip (IC) etc. The historically first dominating the era 1950-1990 were connected on circuits using wires. They were mounted on the circuits boards  (PCBs, we will talk about analytically in next chapter) by passing their wires (leads) through holes and soldering them on the other PCB side that carries the connection copper tracks. Those are the **Through Hole Technology**

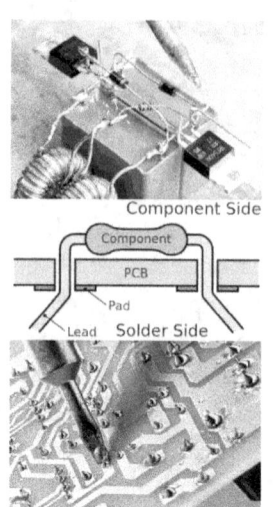

(THT) components, handy for breadboards, easy for hand soldering and making some simple circuits by soldering them together without even using a PCB (image on the left is credited to the great YouTube videos of "Great Scott"). In mass production they are a nightmare. Even in prototyping using PCBs their wires need bending, soldering and then cutting. But they are the beginner's choice as well as the only ones working on breadboards. You should always have some of those for doing quick experiments or small prototypes.

As technology progresses, both for easier mass production and for miniaturizing electronic devices size, **Surface Mount Technology (SMT or SMD)** have come into

the game to dominate more than the 95% of electronic components. 99% at least of all MCUs in the last 20 years are SMD, so are the most important ICs. SMDs do not have wires, but only very sort pins. They are designed to be soldered on the same side of the PCB – on the surface they are mounted on. Not entering SMD technology stops capabilities in almost a stone age, considering mostly the advancements in the last 20+ years in digital electronics. SMDs on the other hand are almost non usable for prototypes, unless a PCB is designed and made for implementing the circuit we want to make. Those PCB prototypes (we will see in the next chapter) enable prototyping of complex projects easily and are cheaply produced on order by many manufacturers (expect cost of around 10$ in Chinese manufacturers for 5-10 pieces of your design). A "middle" way is to use "breakout boards" available for some popular SMD ICs (limited in variety of course) which are small PCBs providing pins to actually convert them into THT that can snap on breadboards.

SIZES OF PASSIVE SMDS

2-pins components come in standardized sizes. Those are the passive components, resistors, capacitors, some coils and others. The sizes you have to remember are those on the left. Their names are from imperial system, 1206 means 0.12" x 0.06" body dimensions etc. Very rarely, metric system names are used which are very different. Those sizes will be well inside your life when you design PCBs. 1206 parts are big, too easy to handle and outdated. 0805 is a moderate size easy enough to handle, not too big for usual projects (projects not space-tight). 0603 for most people will require a little help of a magnifying glass, 0402 are terrifying small and need to have a small microscope for assembling them! Nowadays smartphones have even smaller components (e.g. 0201), available in the market but terribly small to be handled by humans.

1206
0805
0603
0402

RESISTORS

Regardless of choice between THT and SMD technology, when choosing a resistor the following specs must be considered:

Power rating:

In more than 95% of the resistors choice in practical circuits it is not something to even bother about. When V*I may get higher than 0.1W though, it is a consideration.

In the case we do not care, here are the most used sizes we can choose freely from:

2.3 Components: Technologies, sizes and where to find them

Name	Mounting	Usual power rating (Watts)	Dimensions	Description
Axial ¼ W	THT	¼	Body length: ~7mm Diameter: ~2.5mm	The most classic old time resistor.
1206	SMD	¼	3.2mm x 1.6mm	
0805	SMD	0.125 (1/8)	2mm x 1.2mm	Recommended as neither big nor tiny
0603	SMD	0.1 (1/10)	1.55mm x 0.85mm	
0402	SMD	0.06 (1/16)	1mm x 0.5mm	

In case power rating is a concern, size just goes up as the Watts rating goes up. Resistors may work hot (>100°C) when approaching their power rating limit.

Name	Mounting	Usual power rating (Watts)	Dimensions	Description
Axial ½ W	THT	½	Body length: ~9mm Diameter: ~3mm	
Axial 1 W	THT	1	Body length: ~11mm Diameter: ~5mm	Also up to 3W exist
Ceramic or "cement" type 5W	THT	5	Body size~ 21mm x 10mm x 10mm	

			Body size~ 50mm x 10mm x 10mm	It's a rule of physics, the more the power rating, the more the body surface
Ceramic or "cement" type 10W-15W	THT	10-15		
1210	SMD	0.5	3.2mm x 2.5mm	
2512	SMD	1	6.3mm x 3.2mm	About the upper power limit of SMDs

Tolerance:

If we purchase resistors without taking care of their tolerance specs, about 50% of them will be 5% accurate and about 50% will be 1% accurate (in written specifications). Usually the price of 1% tolerance or accuracy resistors is negligibly higher than the 5%. Resistors cost of normal power rating anyway is so low that we will never care (about 1cent of a dollar each in retail dropping the more we buy). If we need for a special application higher accuracy the cost increases as we depart from 1%. Tolerances up to 0.01% are easy to find. If you measure ten 5% resistors, expect to find about nine within less than 2% error. Error changes by soldering them, temperature humidity and operating temperature.

Value:

It shouldn't be possible to find in the market ANY resistor value we may need for our design (e.g. 11234.2 Ω). There should be millions of product models. Instead resistors values are standardized. The higher the accuracy (lower tolerance) the more "analytical" are the values that can be found. In very rare cases when there is not precisely the value we want we may use resistors in series or in parallel to achieve the required one.

5% Standard Values (EIA E24)								
10	11	12	13	15	16	18	20	22	24	27	30
33	36	39	43	47	51	56	62	68	75	82	91

1% Standard Values (EIA E96)											
10.0	10.2	10.5	10.7	11.0	11.3	11.5	11.8	12.1	12.4	12.7	13.0
13.3	13.7	14.0	14.3	14.7	15.0	15.4	15.8	16.2	16.5	16.9	17.4
17.8	18.2	18.7	19.1	19.6	20.0	20.5	21.0	21.5	22.1	22.6	23.2
23.7	24.3	24.9	25.5	26.1	26.7	27.4	28.0	28.7	29.4	30.1	30.9
31.6	32.4	33.2	34.0	34.8	35.7	36.5	37.4	38.3	39.2	40.2	41.2
42.2	43.2	44.2	45.3	46.4	47.5	48.7	49.9	51.1	52.3	53.6	54.9
56.2	57.6	59.0	60.4	61.9	63.4	64.9	66.5	68.1	69.8	71.5	73.2
75.0	76.8	78.7	80.6	82.5	84.5	86.6	88.7	90.9	93.1	95.3	97.6

We can find any multiples or divisions of 10 from those e.g. for 5% there can be 1Ω, 1.1Ω, 1.2Ω, 100K, 110K, 120K etc.

Values easy to find start from 0.01Ω and end at about 10MΩ.

How to read a resistor's value:

So many different values resistors are hard to be organized with labels for each one. The best and guaranteed way is to use a multimeter to measure their value. Most resistors have their value printed on them in a peculiar way. Traditionally THT axial (round) resistors use a color code. They did that for making it possible to read their value from whatever angle you are looking the resistor at. They use 10 colors for each digit, e.g. black is 0, brown is 1 etc. If you "google" "resistor color code" great guides will pop up for this. In SMD resistors as well as in some power THT, it is used to denote with the last digit the power of 10, that is how many zeros to add to the previous digit (e.g. 103 should be 10"+"000 = 10000Ω). A multimeter as previously said is the mostly trusted way to do this. More on measuring instruments in chapter 2.7.

Potentiometers / trimmers:

There are resistors the value of which is can be adjusted manually by

turning a knob or turning a screw head with a screwdriver, all the way from zero to the resistor's maximum value. The knob type are the potentiometers, the screw type are the trimmers (one turn on the left or multi-turn on the right). They come in many sizes. One thing to take care is that they are power rated to less than $1/4^{th}$ of a Watt usually and that the electric power they consume (P=V*I) may vary according to the position they are set (e.g. go too high in very low Ohms settings). Another interesting thing about them is that they provide 3 pins, working as you can see on the left, either as a variable resistor if pins 1 & 3 or 2 & 3 are used, or as a voltage divider if all 3 pins are used. The way it works is actually sliding the wiper metal contact across a resistive material spanning from pin 1 to pin 2.

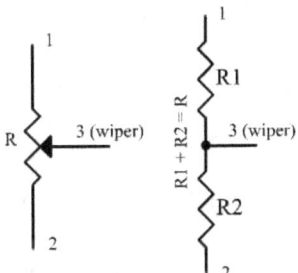

CAPACITORS

In capacitors we care about size and fitting, capacity value and instead of maximum power, maximum permissible voltage.

We will split the capacitors into 3 categories:

1. Less than 200nF
2. Between 200nF and 47uF
3. Over 47uF

In category 1, any capacitor, however small, it is specified to operate to voltages up to 20V, so voltage is very rarely a consideration. In SMDs we have the classic 1206, 0805, 0603 and 0402 sizes. In THTs we have the "radial" body that is like a lentil bean with two wires (leads) coming out (left). Almost all of those (SMD and THT) are made of ceramic material, they are the ceramic capacitors.

In category 2, our job can be done with **ceramic** capacitors and with other kinds as well. In all cases the maximum voltage is our utmost consideration since the more the capacitance, the less goes the maximum allowed voltage for the same size. Ceramic capacitors of this category are also called multi-layer ceramic capacitors. X5R and X7R categories have the best (lowest) tolerance, YxR, degrade their capacity a lot the more the voltage is applied to them. Ceramic capacitors are the best capacitors in performance, they behave almost like ideal capacitors. The next capacitor type of choice is the **electrolytic**. It is a choice of necessity since they are the only ones that can provide capacitance in category 3 or in category 2 in higher voltages (>20V). We have said a few things about those in chapter 1.4. Recapping, (you may see the word "recapping" referring to changing all aged electrolytic capacitors with new!) electrolytic caps have polarity, they explode or die with a puff of smell if voltage is applied to the inverse direction, they have sort life (they are the most common cause of faults in old electronic devices), they have leakage (they self-discharge in about a minute in room temperature) and they have a considerable in series parasitic resistance. About the last, there are more expensive electrolytic capacitors characterized as "low ESR": low Equivalent Series Resistance. Expect about 1Ω -10Ω resistance in most electrolytic capacitors and about 0.2Ω in low ESR ones. Besides electrolytic and ceramics, there are other kinds but not worth mentioning, since they are either outdated technology, or used in very special applications. If you are desperate though for long life or better performance in electrolytic category of capacitors, there is lately a new kind of them called "solid polymer electrolytic capacitors" offering low ESR, long life, but higher cost.

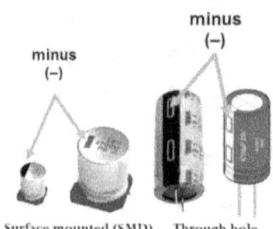

Surface mounted (SMD)   Through hole

Regarding capacitors **value**, they begin from about 1pF (pico Farad!) and end to about 10,000uF (10,000uF/16V is about 20mm diameter X 30mm tall capacitor). There is a class of "ultracapacitors" ranging from 0.5Farad to 10Farads! They are usually rated at 2.5V only but their application is clearly a replacement for a battery rather than a use inside a circuit as a capacitor. Expect a tolerance of about 5% to 30% in capacitors value. Values come at the 5% resistors values (1, 1.1, 1.2, 1.3, 1.5 etc.) on multiples or divisions of 10. We almost never bother of a 20% or even a 50% tolerance in a capacitor. The real capacitance value is also affected by parasitic capacitance. Wires close to each other make some 100s of pF parasitic capacitance. Capacitors connected in parallel (or capacitors parallel to parasitic capacities) add their values ($C_{total} = C_1 + C_2 + \ldots$), like resistors in series. Capacitors in series apply the equation of the resistors in parallel to get the total capacitance.

ACTIVE COMPONENTS: TRANSISTORS AND LINEAR REGULATORS

 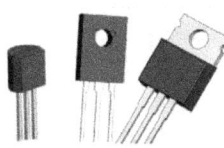

Those are usually 3 pin devices. They both come in THT and in SMD packages. Their size has to do with the maximum power they can dissipate. Those devices take some input electric power (V*I) and output some less, converting a portion of the input electric power to thermal power (an unwanted phenomenon but inevitable). The bigger the dissipated thermal power is the bigger the component body size has to be in order to avoid rising its temperature to dangerous levels. Usually any IC will rest in peace if its temperature goes over about 150°C (300°F) while it is operating. Under some thermal power values (of about 2W) we may choose between SMD or THT technology. Most of their body geometry is standardized in standard body "packages" (named like: SOT23 – SOT 232, TO-92, TO-220 etc.). Since the usual packages of 3 pin components are more than 10 we will not mention them here.

ACTIVE COMPONENTS: ICs

ICs require a connection interface of 3 to hundreds of pins. Most MCUs need from 20 to 100. An extreme far end here is the CPUs of the PCs counting more than 1000 pins. Most ICs need an average of 16 pins. The body size of the chip will depend on how its pins are arranged, how close to each other or sparse they are and how many pins it provides. Sparse pins are easy to handle and solder, close-up pins provide tiny size and / or high pins count. SMD technology goes all the time towards the tiny size. We have to adjust and be able

to "play" with this shortcoming. The pin to pin distance is called **pitch**. On the side of the "easy to use / big in size" ICs there are the THT packages. Almost all of them are called DIP from (Dual inline package) (left). Their pitch is always 0.1 inches, or equally **100mils**,

(1mil is 1/1000th of an inch) or equally 2.54mm. Their pin count is starting from 4 and ends to 40 pins. A 40 pins DIP package IC has a body size dimensions 52mm x 14mm. DIP packages are fitted awesomely in breadboards. Breadboards holes are at 2.54mm / 100mills pitch. Moving on to SMDs, going from easier to handle to smallest in occupying size, we meet:

SOIC packages: Usually ICs with simple functionality, easy to solder, pitch is from 1.27mm to 0.6mm, pin count averagely is 16 pins

QFP packages: Pitch here becomes aggressively small at a standard of 0.5mm. Soldering needs strong magnification glass to inspect if well done. Pin count from 32 to 144pins. A usual package of MCUs(!) with usual ranging from 32 to 100 pins according to GPIOs pins count

QFN packages: Pitch is again at 0.5mm at the most cases, harder to solder since solder is "buried" in a sandwich between the non-exposed pins and the PCBs pads (copper rectangles at the place of each one of the IC's pins). Very usual package (unfortunately) of IC sensors and of MCUs

BGA packages: BGA is for Ball Grid Array. It provides the most space tight solution and the most pain to design PCB properly and to solder. Inspection of soldering needs trial and error or using X-rays machines to see the soldering quality of the internal pins

Well, chips are made to be used by manufactures at more than 99.99% and the rest of the market share is for hobbyists. Manufacturers use robotic assembling machines and need to make small devices, so small pin pitch is the inevitable result. Do not worry. Humans are still in the process, hand manipulation and soldering is achievable to the 99% of the chips in the market. It is not very hard, it needs no expensive equipment. We will visit those techniques on the next chapter. Making small devices after all is a big benefit.

CONNECTORS

There are thousands of connector types, lets focus at the most useful ones. All connectors have a gender, male or female. Well, it is sexism you will surely realize, engineers have named anything that is long, like pins as males and anything that has a hole-shape where pins are inserted as female...

Headers

All classic headers come at 2.54mm (100mils) pitch as single or dual row, male or female with any pin count. Let's see them, as to understand what a "header" is, pictorially:

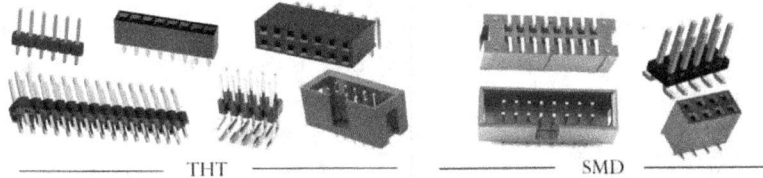

———————— THT ———————————————— SMD ————————

Arduino UNO has single raw female headers to connect things to it. Headers of 50mil (1.27mm) pitch are also used in space tight applications. They are less frequently met though and cost much more. Generally male (pin) headers from China cost around 1$ per 300 pins

Terminals

In more accurate terminology, "terminal block connectors" provide

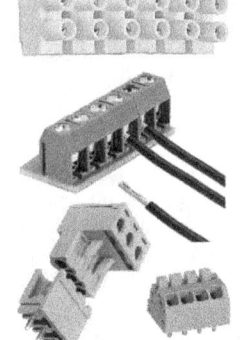

a way to connect / disconnect a cable stripped at its end to another cable or a PCB or a breadboard without soldering / disordering. On the left, from top to bottom, we see some representative terminals: Cable to cable screw terminals, PCB mounted screw terminals, two pieces pluggable type (one piece is permanently mounted on a PCB, the other carrying the cables is detachable) and spring loaded type where instead of using screws we only have to press inwards a spring loaded latching mechanism, insert our cable and release it. The cable then stays latched.

There are many **other** kinds of connectors. Really, they look like never ending, from USB, to power, to waterproof, to board to board signals connecting, to… you name it. Exploring and assessing what is the best for each purpose is up to you for the special needs of a project.

COMPONENTS SHOPPING

Needles to know anything of components and of electronics if you can never get to hold them in your hands. One way to shop is to go to a local electronics component store. Yes, I know what you are thinking, there is probably none. And if you know such a store getting into it to shop components has two issues. One is the very limited stock, for example, if you may have chosen the right MCU of the STMicroelectronics company, that shop may have or may not have that particular one from the around 2000 MCU products of this company only. The other issue is that the process of choosing a component involves studying specifications and parameters. Datasheet reading is not easy when you are on a shop bench. Such shopping could be good for some standard connectors, consumables and tools only.

Thankfully we live in the magic times of e-commerce. Check again paragraph 1.7, section "Introduction To ICs & Components Specifications" for that regarding all electronic components sourcing. Besides components, also about sensors, boards like Arduinos, tools and other equipment there are two worlds. The western world where great but not only presence have the adafruit.com and the spackfun.com and the eastern (Asian) world where great presence (and not only) have the aliexpress.com and the ebay.com.

About China vs West:

a) Distance matters: Depending on your country there may be customs varying in fees and in bureaucracy. Standard mail may also take long or extremely long. Know that courier shipments (e.g. DHL) take less than a week to anywhere on earth (South

Pole is maybe an exception). Shipping fee from many big Chinese companies is descent. Western companies have better "free shipping" options if your bill goes over an amount (only to 10£ in Farnell, around 50$ at Mouser) and courier is usually faster if you live in the western world and usually a little cheaper.

b) Quality: In some **boards, sensors** etc you may expect 1%-3% to reach to you DOA (Dead On Arrival) depending on the specific product and manufacturer, others will be 100% OK. You will still be very happy since they are not 3% cheaper but rather they are >500% cheaper, so just get some spares too. Regarding **components**, my personal so far experience in more than a thousand components (more than a hundred kinds) produced in China from Chinese manufacturers soldered on PCBs had not one single defect.

c) Cost: Expect more than 1/5 in sensors and boards if bought from Asia. About components, you have to see yourself in lcsc.com. Even components like MCUs from western manufacturers are really cheaper as you get the price of ordering 1000 pieces in western distributors in ordering just about 10. Again it has to do with shipping etc. A very important trend in the last decade is the rise of **Chinese semiconductors manufacturers** with descent products at relatively ridiculous prices. A switching type regulator that outputs 5V 1A for example may have lowest price of around 0.5$ in western manufacturers and price of 0.05$ in Chinese manufacturers. Personally I trust those. You might go to western products only if you are on big volume production of a product specified as "cost no object" – "risk 0.0%" only. MCUs are already emerging with great impact companies like Espressif. Other, especially with Chinese datasheets and no software "ecosystem", are not advised to be considered unless you are terribly hungry for lowering a product's cost to about 0.1$ with an extra year of development. Connectors and such stuff are awesomely cheap and reliable also. Once again I will state that I am not affiliated to any company, any country or any politics.

2.4 PCBs, SOLDERING TECHNIQUES AND EQUIPMENT

Having said about components, it is time to see how to assemble them. Making electronic devices is an art. It is not "1+1=2", it is how good it will look, how robust it will be, how quickly it can be done and many other virtues that fall on the artwork category mostly. Let's see the technology and the basic techniques we have to know and use.

From the quickest and "dirtiest" to the more difficult and better performing the most common and not only ways to go are:

1. Connect header to header with DuPont cables
2. Use breadboards (and cables like DuPont)
3. Solder THT components wire to wire, all hanging "on the air"
4. Use a prototyping PCB to solder THD and some SMD components, connecting them directly as well as with wires and cables
5. Design a PCB that implements the connections of our circuit, order it and solder our components on it.

Let's start by explaining what a PCB is, learn the terminology in order to speak about it and then visit ways (4) and (5)

PCBs IN DETAILS

Let's see the most important stuff on a PCB. In this example we have the most common 2 layer PCB that is a PCB having copper traces on top and bottom sides. More than 2 layers have internal layers of connections (copper traces). Complex boards such as PC motherboards or smartphones PCBs may have up to 10 layers or even more. In practical electronics, averagely 1 layer is the 10% of the cases, 2 layers is the 80% of the cases and the other 10% belongs to 4 layers.

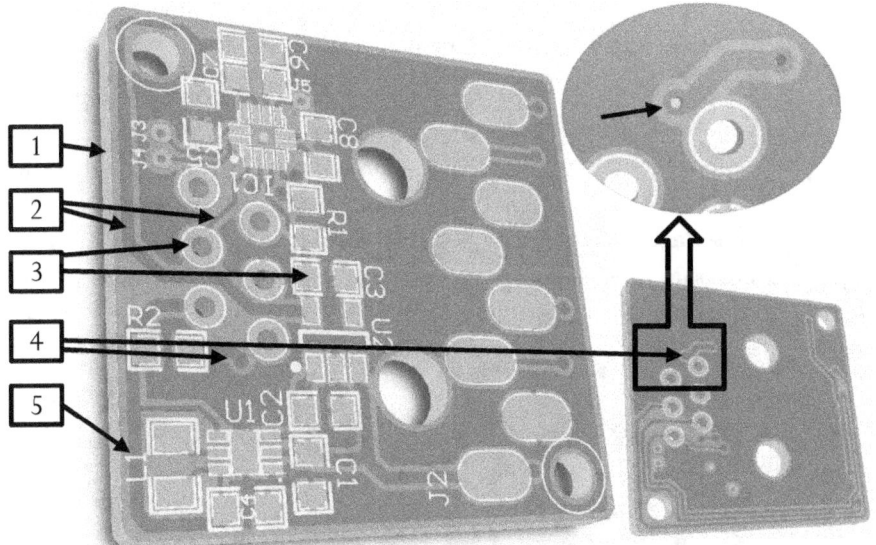

Top layer   **Bottom layer**

1. The PCB **core** is made of a composite material. About more than 98% of PCBs are of FR4 epoxy. It is very hard to bend and to break. The default thickness is 1.6mm
2. Copper traces or **tracks** act as wires. Their thickness is relevant to maximum allowed current and minimum allowed (parasitic) resistance. Same work is done by "flooded" cooper all over an area that acts as an ultra-thick track called **plane** or polygon fill. Copper is usually 0.035mm thick (35um). Since copper is very conductive it takes only 0.25mm track thickness to withstand 1A of current.
3. **Pads**, SMD or THT. Pads are the places where the pins or leads (wires) solder. Through hole pads may be **plated**. Plated pads are like a well with copper wall. They connect tracks from one layer to all other layers (e.g. top to bottom).
4. Same functionality, but without a pad, is done by **via holes**. They are just plated holes. See in (4) a via hole connecting the two points the arrows are indicating.
5. Paints are deposited on top and bottom layers. **Solder mask** is a protective and insulating coating (usually green) that covers all tracks except the pads which have to be exposed for

soldering. **Silkscreen** puts text, labels and other art on our PCB (usually white) to provide information.

Let's now come back to the ways (4) and (5) of our previous list. (4) is about using general purpose prototyping PCBs, which, like the breadboards, are a grid of holes, providing a pad for each hole in order to solder on them THT component's leads or pins. They are called **Padboards** if the pads of each hole are all separated or **Stripboards** if they are grouped in stripes. A way to make circuits on those is like at the following picture:

 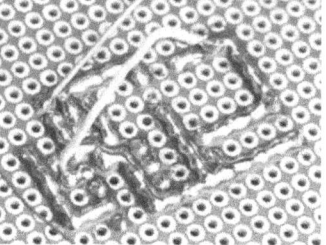

(Images credited to the "GreatScott!" YouTube channel's creator)

Components are placed on one (top) side, on the other (bottom) side they are soldered using their leads or other wires to connect and excessive solder sometimes, like on the right picture. Harder to make than using a breadboard but the resulting circuit is more reliable and robust. About the fifth (and best) way...

DESIGNING AND PRODUCING OUR OWN PCB

Being capable to design our own PCB takes us from moving with a bicycle to moving around with a jet plane. The approach is recommended to circuits with over 10 components or to circuits needed to be made more than twice or to circuits that need to have at least one SMD component that cannot be connected otherwise. PCBs are small, robust and have all professional specs big companies apply to their products. They are also easier and much "happier" to assemble than any other way, provided we have in our hands our PCB. This mission has two parts: Designing it and producing it. Amazingly nowadays both are cheap and fairly quick to do. On designing, the idea is to use software tools that will assist you in making the **PCB layout** that is the geometry of the cooper at each layer, holes solder masks and silkscreen. Those tools, called EDA (Electronic Design Automation), provide ways to design your schematic and then cross over to making the exact connections with the exact components copper pads and tracks forming a real world PCB. The way to become a great PCB designer is quick to begin, it may take a few hours or days for your first simple PCB, but it is long, it may take up to over 6 months of dedication to declare yourself as a pro PCB designer. PCBs are the more complex the more pins that are in the game and that is almost endless in electronics. But practical electronics are usually in the regime or 20 to 200 components and that is manageable well enough. And that is about PCB layout, not schematic design or in other words, electronics design.

There are free to use EDA tools, nice to begin with, like Eagle or Kikad. Going up, there are many advanced professional tools like Altium or Pads coming together with a price to pay to become yours. There are also some online PCB design tools which are free but may lock you in to a PCB producer or components supplier. Going from schematic to layout the process is to fill up all the information for all your schematic components with a **"footprint"** that is its geometry (also 3D in some) of the physical component regarding the pads on the PCB. Next we see a MOSFET that is in a SOT23 SMD package, on its schematic, its 2D footprint consisting of copper pads and other mask layers and its 3D footprint.

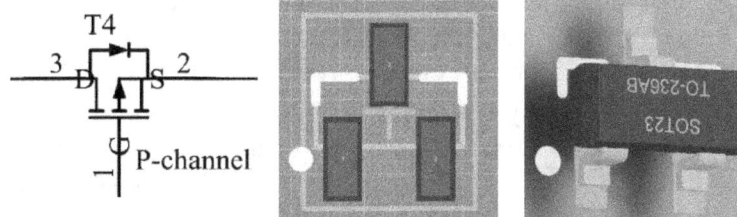

Having schematic made and footprint libraries done, we place the components (their footprints) at the most proper place for each (up to us to decide). Then we go to the process of connecting by drawing the tracks, called **routing**. EDA tools provide auto-routing methods but they usually do not provide as good results as taking the patience and make them all by hand.

Design considerations to be taken are many, others are electrical like the resistance of a track, most are manufacturing, regarding whether a PCB manufacturer can produce that or not. The most important such **"design rules"** are the minimum track width and track clearance, having to do with the feasibility in resolution or detail in the production of a PCB. They are usually measured in mils (mili-inch, 1mil=0.001"=0.0254mm). Usual constrains of most PCB manufacturers are 6mil minimum width / 6mil minimum clearance for standard cost. They go lower with extra cost, down to about 3mils but that would be rarely needed. Holes size should also not be less than 0.3mm and finally, the copper around a hole pad should not be too thin (<6mils) and holes should not be too close to each other. EDA programs have "design rules" settings to take care of all those but the designer should be aware to set them up and to take care for them.

Having our PCB layout done, we need to convert it from information to reality. In the old days (15 years ago) either we made it ourselves or we spent a lot of money (200$+) and waited for more than 10 days to have others make it for us. In this fairytale age we are living now, prototype PCB manufacturers in China can produce it for us asking only for about 5$ total for 5 pieces (sizes up to double a credit card size), one to two days to manufacture and very reasonable courier shipping costs. Examples of such factories are PCBWay, JLCPCB and many others. Western world has it factories also but you have to make a price and delivery time comparison to see if there is meaning for that. The biggest personal experience I have is getting about 40 different orders of 20 pieces average (for each order) of fairly demanding designs of PCBs from PCBWay, all without one problem and on time. Needless to say that I am not affiliated to them. You should check at any manufacturer the technical capabilities he offers and try to fill its automatic quotation form. This information will make clear your "design rules". Note that there is a standard file format for PCBs layouts called **"Gerber"** files and **"NC drill"** files, exported by all EDA tools.

This discussion about PCBs is not intended to make you a PCB designer, but only to teach you the terminology and where to begin from if you ever embark to the journey of a project of yours that needs it.

SOLDERING TECHNIQUES

Soldering is an art. It is achievable to a very great level by most people. I believe more than 95% of the people can make it to a very high level and the rest can achieve medium to good level. All you need is surprisingly, not steady hands, it is good close up vision (glasses guarantee top level specs on that) and love and eager to make what you are about to assemble.

We may categorize soldering in two kinds. Soldering THT (through hole) and big and easy SMD components and soldering SMD fine pitch components. For the first kind it is estimated that you should train on real projects for about 1-4 hours to feel like you go for it. The other kind is twice as hard and may cost a few burned components also.

A picture is 1000 words they say. How about 16000 pictures, one after another, in a 11'08" video? Soldering contains motion also, so let's not do it with words, nor with pictures, open your YouTube and search for

great scott solder

click on the video titled:

 *"How to Solder properly || Through-hole (THT) & Surface-mount (SMD)"*

and enjoy those 16000 frames.

Videos are the way to learn about soldering. Count those 12 minutes in the time of reading this book. In the spirit of this book, that was the introduction, there is way more to go, learning and experiencing.

Companies like PCBWay offer also **PCB Assembly services** (PCBA). Having seen on this video how the solder paste works, imagine applying the solder paste only on the pads with one move using a stencil and a spatula and then using robots that do automatic placement of components (automatic pick and place machines). This is about mass manufacturing. Nowadays many companies offer very low prices for automatic assembly of your boards by such precise machines, even for low count of boards e.g. 10 pieces only. Cost may be around 100-300$ fixed fee plus about 3$ per board of around 100 components each. That makes you an electronic product manufacturer who owns no fabrication facilities (fabless) and that is really an awesome potential. Special information for those automated machines (pick & place files) as well as detailed bill of materials (BOM) is generated by EDA tools. Design and soldering experience is needed to do this though, your first projects should better be assembled by yourself.

EQUIPMENT BUDGET

Back in 80's and 90's a well-equipped electronics lab for developing or just playing with MCUs or CPUs costed well over 5000$. Nowadays we live in heaven, the same professional level functionality in **equipment** that includes measuring instruments, tools and programming/debugging equipment **may cost less than 200$** provided you have a PC, desktop or laptop. We may categorize the cost of an electronics laboratory in four levels. The spectrum is continuous, you will not step from one to another such subjective level but be anywhere in between.

Level 1: Low-end but well-chosen equipment in tools and instruments may cost less than 100$. Some simple enough Arduino projects may need no equipment at all, just the components, the Arduino, the PC and a few DuPont cables.

Level 2: Medium, 95%-99% functional for developing anything with Arduinos and MCUs in custom PCBs may cost around 500$!

Level 3: Very good, >99% functional for easy prototype assembly or repairing of any small SMD size components costing around 1500$

Level 4: High-end equipment covering a missing <1% (such as high frequency, ultra-high accuracy instruments) if ever needed in projects involved) costing in the region of 5k$ to 100k$.

A laboratory for the sake of practicality, should also be equipped with a stock of common use components and consumables (some of those purchased for the sole purpose to play with them once for acquiring experience, knowledge, curiosity satisfaction) over at least the half of the equipment cost.

So, it's an "anyone can make" job.

MOST IMPORTANT EQUIPMENT – MEASURING INSTRUMENTS NOT INCLUDED:

A shorted by priority list follows to all you need in your lab, leaving out the measuring and debugging instruments (ultimately important) we will talk about in chapter 2.7. Do not take all that follows by word, prioritizing is subjective and mostly refers to where to put more money in case the budget is very constrained. It is not about having only the first ones and nothing of the ones following. Budget is mostly referred to getting things from China or Ebay.com.

1. Soldering station: A temperature controlled soldering iron with a variety of soldering tips. In choosing such, the 90% of the weight in the decision factors is what variety and quality of tips 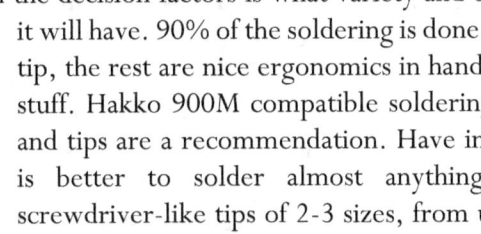 it will have. 90% of the soldering is done with the iron tip, the rest are nice ergonomics in handling and such stuff. Hakko 900M compatible soldering iron "pens" and tips are a recommendation. Have in mind that it is better to solder almost anything using flat, screwdriver-like tips of 2-3 sizes, from ultra-small to quite big. Keep also in mind that different tips make different temperature at their end point, so do not believe the temperature your station is set to have. Higher temperature than required wears off your tip sooner and most importantly, may damage your components like ICs and transistors. For low  budget take into consideration some "pen-only" with all electronics integrated. For stations a turning knob has more usability than up-down buttons provided it is not 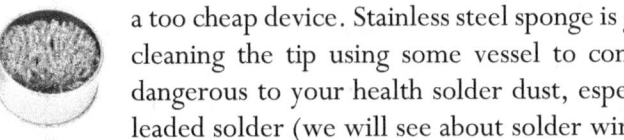 a too cheap device. Stainless steel sponge is great for cleaning the tip using some vessel to contain the dangerous to your health solder dust, especially of leaded solder (we will see about solder wire next).
Budget: Level 1: 10-40, Level 2+: >60$
2. You have to have great close-up eyesight (presbyopia with no glasses is very bad, myopia is a bliss). If not use glasses only for electronics as to achieve close-up eyesight at 100%. Regardless eyesight performance, in order to see smaller objects (also like

 bad soldering) magnifying glasses like the ones on the left are recommended (with lenses shape as seen in the photo) with strong but not too strong magnification. "Sherlock Holmes" style magnifying glasses, if used instead of wearable glasses, should be small and "sort focus", the bigger diameter they have, the less they magnify! Any magnifying glass should not be used for prolonged times. Budget: 4-30. For Level 3 consider a 3D optical microscope with the <u>lowest</u> possible magnification + greatest distance from objects viewed. Camera & screen approach is not recommended.

3. Cutters: Have two at least, a fine and a bigger, with as good  quality as possible. Stripping cable ends easily with your cutter is what makes quality pay off. The fine cutter should have the form of the one on the left. Obey the maximum diameter of wires written on its handles always and never drop it from your desk. If you cut anything harder or you drop it with its tip hitting the floor go for another one, so have at least one backup. Big cutters of Knipex German brand are recommended if you can put another 30$ there. Budget: Level 1: 5$ Level 2+: >20$

4. Tweezers: You cannot handle any SMD component with your fingers. Using SMD turns you into an alien creature with one hand of two metal fingers that is your metallic tweezer. You should allocate some budget to it for top quality, like titanium or other super metals. Also have more than one and more than one size (at least a thick for harder jobs and a very fine – thus sensitive – for SMDs). Budget: Level 1:>5$, Level 2+: >30$

5. Pliers, blade cutters and other general purpose tools (many you may have already)

6. Screwdrivers: Silly to talk about those but have in mind that having 2-4 most common sizes of very good quality like the French Facom brand pay off, mostly in pleasure. My guess is that you already have screwdrivers.

7. ESD protection: ESD stands for Electro-Static Discharge. This is about tiny sparks we do with our hands when we touch objects while our cloths have accumulated (static) electric charges. The

voltage produced that way is awesomely high, over 3000V! The current is just tiny. This damages in permanent failure many ICs and some MOSFETs. But how often does it occur? It depends a lot on environmental conditions like humidity and varies in severity from place to place (e.g. may be very rare in Greece where I live and have done no such damage yet, it may be more dangerous to dry places like USA Texas). Generally it is a precaution that even if it is not taken, most of the times no problem occurs. The way to protect from static electricity is to have electric conduct of both ourselves and the components we handle to the same voltage, the ground (here the term ground refers literally to the ground of the earth). Packaging with material of a very little conductivity (ESD safe bags) is also helping on this a lot by sort circuiting all IC's pins to the same voltage. To be ESD safe we need to connect our conductive human body to the same voltage with our bench and instruments. The main equipment is an ESD mat on our bench to work on (costing from 6$) and an ESD bracelet (costing from 1$) that we have to wear always in our hand. Both have a cable (some cordless ones are probably just scam) that we should connect with each other and with the electric grid's ground or earth point. Choose the most "rigid" pad you can find.

8. Hot air rework station: They are the only means of de-soldering 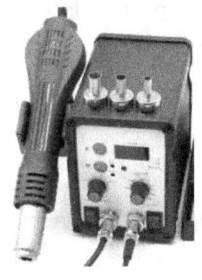SMD chips, and soldering many kinds like QFNs and many other SMD packages, including almost all quick works using solder paste. So it is an SMDs must have. Many come with a soldering iron also. Do not choose some old technology of those with an air pump inside the base station box. Newer ones contain a blower inside their hot air gun that connects to the station with a cable only (not an air hose). They are quieter and have better flow control. Using such a tool requires some YouTubing to learn the skills. Notice that the temperature of the air depends a lot to the distance your nozzle is and the air flow rate, so the air temperature setting is to be taken in very little consideration, burning your ICs by overheating will be your constant fear on this job. (*Image is*

credited to Yarboly Authorize Specialty Store in Aliexpress.com) Budget: Level 1: 20-30$, Level 2+: >40$

9. Cable stripper: There are many kinds. Most cable-end stripping, in practice is done with your good cutter. Budget: 5$ -30$

10. Hot glue gun: For any ad-hoc mounting especially for wires. Take care not to apply hot glue on operating ICs since it is hot enough (>150°C) to damage them. Is OK for powered off electronics. If you burn your hands toothpaste is a great relief.

11. Hot plate: Hot plates provide a surface with fine temperature control (practically from 100°C to 300°C) that heats your board at its bottom. They alone or with help from hot air from above can melt the solder of SMD parts either for soldering or for de-soldering. A very great asset for QFN chips. Older type IR heating "pre-heater" devices do not perform as well. Recommended for level 2+ of equipment, costing from 100$ to 300$ where the highest portion of the cost is the shipping due to their size.

12. Reflow oven: A small oven where you place your board with the SMD components and solder paste in order to melt the solder paste properly and deliver your board with all parts soldered. It achieves "temperature profiling" that is rising the temperature by specific rates and holding it to specific points for programmed time periods according to what solder paste likes most for soldering well. Cost begins at around 300$ and it is at level 3 of our budget categories.

MOST IMPORTANT CONSUMABLES:

Here is a list of what should be always within the reach of your hands, again sorted according to importance:

1. DuPont cables: Form all four kinds, more than 50 of each, especially of female to female and female to male. Take care to have as thick core as possible (less AWG). Cost: around 5$ at least.

2. **Solder wire**: It looks like a trivial thing but its quality is the  most important factor in soldering! First of all, there are two kinds: **Lead-free** and **leaded**. Leaded is the oldest and cheaper, it comprises of lead (Pb) at about 40% and the rest 60% is tin (Sn). It melts at 188°C (370°F) that is much lower than lead-free. Generally it is "worked" better than the lead-free BUT it is not healthy. Lead is not good to come in conduct with your skin for prolonged times. Lead-free is the healthy solution to you composed of various combinations of tin, copper, silver and antimony, but with a higher melting point, 217°C (422°F), it is more dangerous to burn some components if care is not taken. The metal alloy itself, of both kinds, when melting, takes a plastering form and is hard to be deposited or to join with the pins, wires or pads we are soldering. There is a second ingredient needed to solder, the **flux**. Flux makes the melted solder flow non-viscously, like a drop of water, makes the joining to the soldered metals (copper etc.) a lot easier and forms a coating around the finished soldering that prevents from oxidization. The flux is contained inside the solder wire. Flux makes the smoke while solder is melt. Its fumes, especially if it is from natural rosin are harmful to asthmatic people at least. Yea... natural products are not good for your health. Generally fumes are not toxic but you should not inhale them a lot. Use a very low speed current of air flowing horizontally above the board that you solder (with a small fan located in some distance, air flow has to be small otherwise it will cool down your soldering iron a lot) in order to take the fumes away from your face (that is usually right above) and some ventilation, or professional air sucking equipment if you are in production of electronic boards. Flux also leaves a residue around your soldering that absorbs humidity and becomes conductive after a few months making the most terrible sort circuits since they will one day come out of nowhere. There are two flux categories, **no-clean** and others, usually unspecified and cheap. Amazingly no-clean is clean!! They set that confusing name to inform that it does not require cleaning after soldering, it will not form parasitic resistances after time.

Another specification for our solder wire is how much flux it contains. We need to have lots of no-clean flux (the most possible actually) that is in around 3% **concentration**. That makes all soldering much more easy to do. You will find great info and great products in kester.com web site. A last thing is about the **diameter** of the solder. 0.5mm diameter is fine for all THTs and all, even the smallest SMDs. Prepare to spend around 50$ for a well selected roll of 0.5Kg of solder wire that will serve you for some years of working on electronics. Avoid solder wires of wrong diameter, without no-clean flux or without detailed specifications.

3. Spare cables of various diameters, from very thin to about 1.5mm copper diamter. Especially it is recommended to have some rolls or long pieces of different color, with wire (core) diameter around 24AWG (equal to 0.22mm^2) and some ultra-thin. Single core (solid copper wire) breaks if bent many times, prefer "multi-core". Cost here may be around 5$ at least. In worst case strip-off some network cable.

4. Spare components of various kinds such as: LEDs (more than 10), THT resistors, more than 50 (too cheap) of each of 1Ω, 10Ω, 100 Ω, 1K, 10K, 100K, THT capacitors, more than 50 of 100nF, 10uF, some of 10pF and some bigger electrolytic. The more kinds and quantity you have, the better. Cost may be as low as 5$ but around 20$+ is recommended. If you are to design and make SMDs circuits, SMD assortment kits of resistors and capacitors are a must. Recommended are ones containing more than 50 values at 50 pieces from each at least, organized (assorted) per value. You should need 0805, 0603 sizes at least. For levels 1 and 2 you may get such kits starting at 3$ in Ebay each, for level 3+ consider some nice practical booklets and assortment kits of lots of component kinds. When you need to experiment with something, you cannot go shopping and wait some days e.g. for your slightly bigger resistor to arrive.

5. Breadboards and prototype boards. Allocate more than 5$ for those.

6. No-clean rework flux: Extra flux in a practical syringe dispenser, essential for SMD ICs soldering and unsoldering. A

must for working with SMDs! 5mls of volume should do your job for around 100 chips.  Old school electronics guys used a small tin of flux to dip their iron tip inside. No, this is pointless added to the fact that such fluxes have terrible quality and are not "no-clean". "Rework" kind of flux means it stays there for longer time while heat is applied, in contrast to water based fluxes. Costs around 10$.

7. Flux removal solvent: Even no-clean flux has to be removed since it is annoyingly sticky and it does not allow for good optical inspection of our soldering. Either use special sprays labeled for this purpose or use isopropyl alcohol. Any of those is a must have on your bench. Remove the majority of the flux while diluted to the solvent with a paper towel or better with expendable micro-fiber cloths. The later do not leave any annoying paper fibers behind. In some cases, re-apply new solvent and use a toothbrush assigned for that purpose to make your flux-covered area ultra-clean (provided your toothbrush is kept relatively clean also). Use ventilation in your space. Cost starts around 4$, beware of shipping of chemicals issues and cost from abroad.

8. Solder removal wick: A special copper cable with embedded  flux (make your own if you don't own it) that sucks all solder from a solder joint when heated up on it with a soldering iron (high temperature does better job). Cost starts from 1$

9. Adhesive tapes: Plastic insulating tape for providing proper electrical insulation to bare wires, paper tape for easy closing of components bags and for mounting things in a cleaner way, does not leave glue behind when removed. "Capton" heat resistant tape that withstands even the temperature of your hot air and leaves no glue behind also. Cost from 1$ to 10$

10. Heat shrinking tubes and wire ties for greater look of cabling work. Cost from 4$

11. Solder paste, definitely no-clean flux. Though you can live without paste even for the finest SMDs. Its place is in your

refrigerator, safely out of reach of children or people with limited cognition. Costing around 20$ in practical syringe dispensers.
12. Last but not least: "Consumable" devices. Even they are not consumables in nature, it is not bad to have an in-house collection of a few Arduinos of various kinds, and a few sensors and peripheral boards in general of various kinds that are big candidates to be used sometime like temperature sensors, relay boards etc. Since your Arduino may be damaged quite easily from mistakes, at least one more should be awaiting "on the bench" to come into the game when needed.

RECOMMENDED LAB LAYOUT

Since hardware with no software is now rare and since hardware itself is designed in computers and last, since all knowledge is either in .pdf format or is "googled" you cannot create electronics nor repair them away from a PC. So a desktop PC or a laptop (not a tablet) should be no further than 1 meter from the place all your tools and instruments are. Note that if you take this a little seriously, a PC having at least two monitors for efficient designing and software development should be there. Fortunately a good bench / desk with all level 2 or 3 of equipment on it and sufficient working area can be as sort as 1 to 1.5meters in length + the space for the computer.

2.5 FURTHER "MUST-HAVE" KNOWLEDGE ON PASSIVE COMPONENTS AND SIGNALS

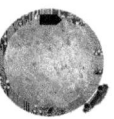

Previous two chapters were easy to read, just describing physical – hand grasping objects. Electronics, like it or not, are based on science (👍👍👍), that is physics with the assistance of a few mathematics. This and the next two chapters will be hard core electronics deeper concepts, all 100% practical and useful every day. In this chapter we will dig in to how diodes and capacitors behave and gain some background to understand important information in most component's datasheets.

FUNCTIONS, LINEAR AND LOG PLOTTING

Skip this paragraph if you know by the title what those are. But for anyone who is not on mathematics or on engineering, let's not leave any gap behind since understanding and knowing are two completely different things.

Most of the practical mathematics are **functions**. Simplest functions (the 99% of the ones we meet) are "machines" which take an input of one value (that is a *variable*) and output something depending only on how much that input variable is. For example a function may be the $f(x) = 2*x$ that has x for its input variable. Feed it with 10 it will output 20 always, every day, no matter the weather. We understand a function's "functionality" as a picture by

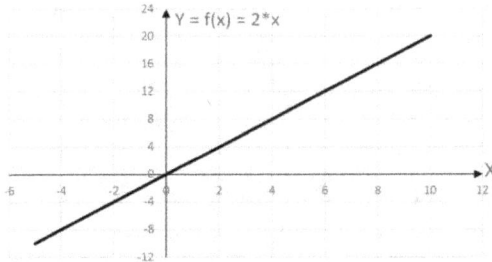

drawing a plot on two axis, the horizontal is any input value range we are interested about, the vertical is the result of each of those. Surely you see many such plots in your ordinary lives, usually having for input variable the time and for output a measurement or the predicted value of something, that easily displays if something is ascending or descending. We may call a function **"something"** vs **"something else"** e.g. Apple's stock value vs time (always output versus input).

Notice the numbers on the two **axes** in the previous diagram. Those are **linear axes**, increasing their values by the same rate throughout their length. If we choose some length e.g. 1cm in our paper, X's value increases by something, say for example by 3.2 units. Taken that 1cm at any point of the axis we see the same increment (3.2 units) no matter how left or right we are.

Sometimes in engineering we need to display very small and very big changes all together. We have logarithmic or **log axes** for that. Unfortunately they work for positive values only and they cannot include the zero value. They "zoom in" in great detail the closer to zero we are that is, the more left or down we are and they "zoom out" to big numbers the more right or up we are. Here is one:

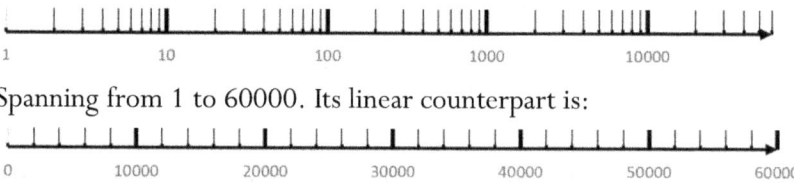

Spanning from 1 to 60000. Its linear counterpart is:

The thinner lines of our log axis are values: 1,2,3,4,5,6,7,8,9, (10),20,30,40,50,…,(100),200,300,… while on the linear axis they are: 0, 2000,4000,6000,8000,(10000),12000,… Compare the 1-100 area of the logarithmic to that of the linear. Log axes have the mathematical magic that if we do what we did our previous linear axis with the 1cm area moving it left or right, we see that in any place we set it the value increases by the same **multiple**, e.g. it increases by 4x! (Easy to see that on the 10x increment). Let's remind again that zeroes counting on numbers is easily displayed as

10^{-3}	0.001
10^{-2}	0.01
10^{-1}	0.1
10^{0}	1
10^{1}	10
10^{2}	100
10^{3}	1000

powers of 10 (orders of magnitude in physics slang) as in the matrix on the left. It is very usual to see log axes spanning e.g. from 10^{-6} to 10^{2}. All are positive numbers, not containing the zero. Enough of mathematics tools, lets jump to the real resistors and diodes behavior.

RESISTORS

Let's plot the current flowing versus the voltage applied on a resistor. Note once more, that voltage is usually what is set on a component or a sub-circuit and current is the result of its behavior. Available are voltage sources in the most of the cases. Current sources (making current have a specific / set value) exist as well but we will see those only in transistors, operational amplifiers and in very few more cases, all considered quite rare and tricky. So, having voltage as our input variable, Ohms low is I = V/R (higher resistance, less current, higher voltage, more current). Plotting on the left 0.1Ω, 1Ω, 2Ω and 10Ω all on the same diagram as to have immediate comparison, we see that the less the resistance the steeper is the slope of the line, the more it is the less is the slope. Let's start from the plot of 1Ω. Two Volts make 2Amps, 10V make 10A etc.., going higher in resistance, 10V will make less and less Amps up to the point of reaching infinite resistance were 10V and any other (input) voltage results in 0A, that is a horizontal line coinciding with the V axis. Infinite resistance is the open circuit. Likewise, zero resistance that is the sort circuit (with an ideal wire drawn on our schematics) makes an I vs V diagram of a line on the vertical (I) axis. In that case, any, however small voltage will produce infinite current. Don't start to do much philosophy on that, there are no real 0Ω resistors as we previously show how much the parasitic resistance of copper wires is. Moreover we show that all voltage sources have some internal resistance, but rising current too high usually results in circuit death with or without smoke.

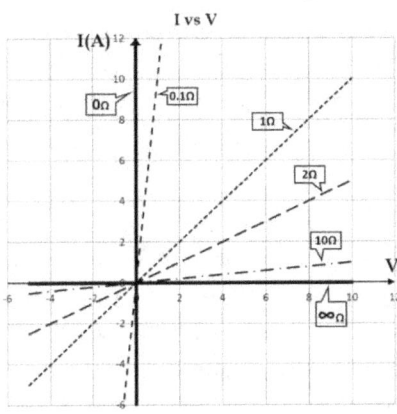

Diodes

Let's apply voltage of various values, positive and negative (reverse polarity than that shown) on a diode, assuming it behaves ideally.

Such an ideal diode should behave as shown on the diagram on the left. Open circuit on negative voltage applied (called reverse voltage), ideal sort circuit (zero resistance) to any positive voltage applied (called forward voltage), behaving like an ideal wire. Unfortunately real world components are not there yet. Real diodes have many performance limitations, the main of those are:

Maximum current they can handle (on positive voltages, or "forward biased"), a little leakage current on reverse voltages (behaving like resistors of some $M\Omega$ value), maximum reverse voltage they can handle and as we will see next, a "forward voltage drop". Here are the real I vs V diagrams of some real diodes, SS14 and some others "Schottky" type and 1N400x (x is 1 to 7) classic diodes (or silicon rectifiers kind), both handling up to 1Amp current.

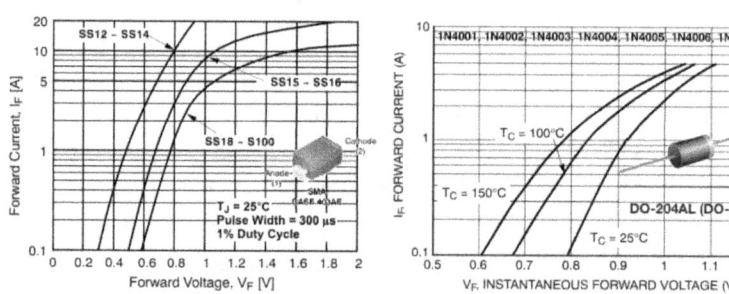

1A diodes are medium in size, smallest are around 200mA maximum current, biggest practical are around 5A. Diagrams above show current over 1A that they can handle only on sort durations. Note the logarithmic vertical axis. Now notice the main difference between the real and the ideal diodes. Instead of "turning on" at 0V or at 0.000001V they start to act like a sort circuit after a voltage

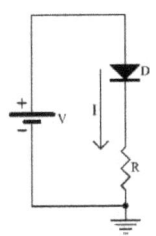

called V_F or $V_{FORWARD}$. If current of 200mA flows on the circuit on the left, SS14 will need around 0.32V to "turn on", SS15, around 0.52V and 1N4007 around 0.83V. That voltage will be the voltage the diode has across its pins when such current is flowing through it. So, in our circuit on the left, if we have an ideal diode for D, the voltage source equal to 5V, the resistor equal to 25 Ohms, voltage across D should be 0V, voltage across the resistor should be 5V-0V = 5V and the current should be I = V/R = 5/25 = 0.2A. Having the 1N4007 for the diode, current should be again around 200mA (a little less as we will see but that is not significant) but the voltage across the D now should be around 0.8V, so the voltage across R should be 5V-0.8V = 4.2V ! So a Diode makes a voltage drop, depending on its specs (which one it is) and on the current that flows through it. Of course that is not wanted. Diodes of "Schottky" type are the best at this with a tradeoff of more leakage current in reverse voltages. Also note that many <200mA applications are covered well with the most classic diode (silicon rectifier kind), 1N4148. You should always choose a diode easy to find on the market since there are a lot to choose from for any set of specifications. Octotpart.com and other alike sites help on availability check and that goes for all components selection. Let's now visit some very fancy diodes, Led Emitting Diodes.

LEDs

The more current flows through an LED the more light it emits. That is for most cases proportional, double current make double (not triple) light. Current and thus light output have an upper limit, for most LEDs purposed for indication of something (rather than lighting the surrounding environment) it usually is 20mA. So, 20mA gives maximum light, 10mA gives 50%, 1mA gives 5%, 0mA gives zero.

Let's see the forward voltage drop of some typical LEDs, a red and a blue one. Notice though that each part number of an LED has its own such diagrams in its datasheet which may differ.

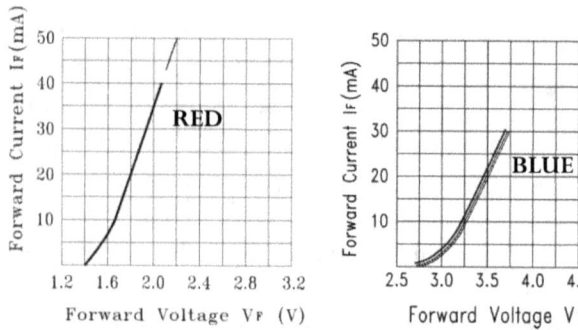

Notice that this drop is a lot bigger than the diodes have and therefore it plays a more important parameter in designing. We will come to that next with a designing example.

Other parameters important to LEDs are: Color (it may also be referred as wavelength in nanometers), luminous intensity expressed in milicandelas (mcd), 100mcd is a bright enough to feel it too bright when we stare at it (a very bright one is power economical since we achieve the light level needed with less current), its maximum forward current and its overall shape and lens if any.

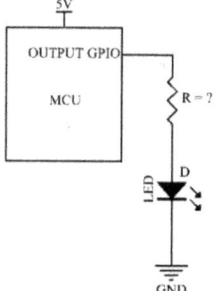

Let's design... In the circuit on the left assume we have a blue LED and we require it to shine with its maximum intensity. What resistor should we choose? Assume the LED has max forward current of 20mA. We should go with 15mA to be sure there is always enough margin. When the current of the LED is equal to 15mA, the voltage across its pins is around 3.3V according to it's I/V diagram. Assuming the GPIO output of the MCU works ideally (like a switch, no in series resistance) when it is set to "high" state it will be connected to Vcc that is 5V. So R will have 5V on its one pin and 3.3V on the other, making the voltage across it 5V - 3.3V = 1.7V. Of the 3 elements of Ohm's law we have voltage and current, so resistance is R=V/I = 1.7V/0.015A = 113Ω. Taking into account that the GPIO sub-circuit may have around 30Ω in series resistance (parasitic) we may either chose an 82Ω since a margin is already left or play safer with an100Ω. We should not care about how many

Watts the resistor is since the power dissipated is V*I=1.7*0.015=0.025W

Note that this is near the limit a GPIO pin can source, for higher current the help of a MOSFET should be needed. Without a resistor in series to an LED, in the red one for example, 1.8V makes it shine the most, 1.7V makes it shine to half and 1.85V takes it out of its operating range. That's why an in series resistor is always needed. With an in series resistor we are very tolerant on both its value and most importantly on the voltage we apply. In our previous example, raising 5V to 5.5V should make the LED current rise by about 3mA.

CALCULATIONS ON CAPACITORS AND BATTERIES

In Chapter 1.4 we pictured the capacitors as "charge reservoirs" like the gas reservoir of a car. Let's introduce how to make the most basic calculations about "how much" millage such reservoir provides as to pick the right capacitor for our jobs. But before that, let's talk about such calculations on batteries.

A **Battery's** front-end specifications are its nominal voltage and its capacity in Ampere-Hours. The Ampere-hour (AH) unit is actually the product of **current** x **time** (in units A and hours) that remains **constant** and is a measure of the amount of the energy that can be provided in total before our battery depletes. One AA alkaline battery that is 1.5V, 2AH can provide 1A for 2 hours (current x time = 2AH) or 0.1A for 20 hours (our product is still the same) or 200uA for about one year or… That's the concept of the capacity of a battery. Note that in the real world we cannot have in our AA battery 10A for 0.1H for example, when the current goes very high our AH product drops.

A **capacitor** is similar to that, but is almost ideal in providing high current and recharging trillions of times without wearing off. The measure of its "gasoline tank" volume is not measured in AH. The measure of capacitors' capacity itself (Farad units) is defined as: One Farad is 1 Coulomb of charge stored inside our capacitor (remember current is $I_{(Amperes)} = \dfrac{Q_{(Coulombs)}}{t_{(sec)}}$) for 1 Volt of voltage applied across our capacitor's pins. If 1V is applied on our capacitor, electric charge is stored inside it, the more that charge is, the bigger the capacity. For 1V if the stored charge is 1 Coulomb, our cap is 1 Farad. Doubling voltage doubles the charge, e.g. in our 1F capacitor, 5V will make it contain 5 Coulombs charge. At discharging our 1F cap, if 1 Coulomb flows out, its voltage will drop by 1V (e.g. in the previous case, its voltage will drop from 5V to 4V). That 1 Coulomb, flowing out in our circuit, will provide 1A for 1 second or 1mA for 1000sec or… any constant current x time product (in A x sec this time, we may call this Ampere-seconds like AH) but for dropping by 1V, not for depleting. A 1uF capacitor (0.000001F) has 1uA-secs for dropping by 1V. It can so provide 1A for 1usec for dropping 1V or 0.01A for 1usec for dropping only by 0.01V or 0.01A for 10usec for dropping 0.1V or…. You name it by keeping the **current** x **time** product **equal** to its **capacity** x **voltage dropping** product. Same applies for charging the capacitor, voltage then rises by the same amount.

$$I \cdot t = C \cdot \Delta V \quad \text{or} \quad \Delta V = \frac{I \cdot t}{C}$$

In our real world of components, the maximum current a capacitor can provide (or be provided) depends only on the ESR, the Equivalent Series Resistance we mentioned in 2.3 that every capacitor has. Ceramic capacitors (usual capacitors below 10uF) have very few mΩ, so the current they may provide can be huge (though very sort lasting, in the region of nsecs only). ESR is a concern for electrolytic caps as we already discussed (going up to a few Ohms), in general the smaller the capacity, the smaller is the ESR.

Digital ICs consume current every time they change a state in them. That consumption is in spikes of sort timed pulses (with duration of around 1nsec). Placing a capacitor near them (usually a 100nF) can serve that high current demand in those sort durations. 100nF may provide a pulse of 1A for 1nsec having only 10mV voltage drop. Same concept applies to batteries, especially the very small sized, power supplies connected over long wires etc. where a near-by capacitor supplies all sort-term current demands keeping the supply voltage stable enough.

MORE ON SIGNALS

Hereafter in our diagrams and concepts we introduce time. This will introduce other great concepts such as frequency. We will also get used to the fast time ticking of electronics, where 1milisecond sometimes is like an eternity. On chapter 2.7 we will see how we actually see those graphs on our oscilloscope's screen.

Square wave:

Let's use an MCU GPIO set as output and write a program like:

```
void loop()
{
    digitalWrite(GPIOnumb,1);
    delay(50);
    digitalWrite(GPIOnumb,0);
    delay(50);
}
```

The GPIO, provided the MCU is supplied by 3.3V for Vcc, will start changing over time as in the following time-axis diagram:

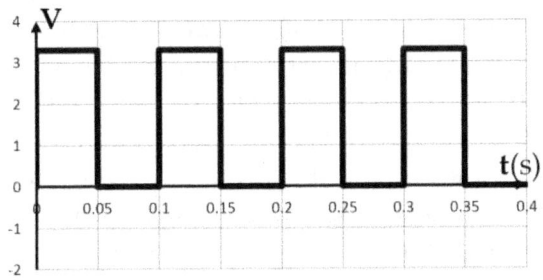

Diagrams where the input variable (horizontal axis) is time show how something evolves – changes. You may meet the term "transient response" for this.

Our diagram or graph shows a periodic change. That horizontal-vertical-horizontal-vertical periodic signal is called square signal or square waveform (or square wave). Some very important terminology on this follows:

Period: the time duration (in **seconds**) of the periodic phenomenon (one cycle). In our case it is 0.1sec (not 0.05!). Also note that one period is also the time from 0.02 to 0.12sec or any other 0.1 sec interval.

Frequency: How many periods (cycles) occur every passing second. Yes, it is the reciprocal of period, $f = \dfrac{1}{period}$ measured in units $\frac{1}{sec}$ or same thing written in another way, sec^{-1} or (same thing) Hertz or **Hz** after Mr. Hertz a pioneer in radio waves around 1880. In our example it is 10Hz.

Peak to peak amplitude: The difference from top to bottom in the vertical axis. In our diagram it is voltage and is $3.3V_{p-p}$.

Amplitude: This is confusing. It is the half of the peak to peak amplitude. In our case it is 1.65V.

Duty cycle: Only in square waveform: The **ratio (%)** of time that is high to the time of one period, in our case it has been 50%. On the diagram on the left it is 10% (yes that is a square wave too). Adjusting the duty cycle of a square signal is a process called Pulse Width Modulation or

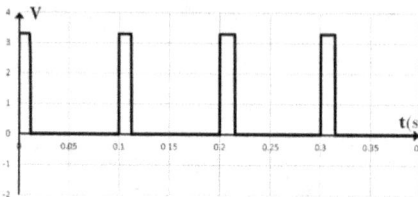

PWM. The average value of a square wave that is zero when low equals to its peak (high level) value by the duty cycle ratio. In our case of 10% duty cycle signal of $3.3V_{p-p}$ it is the 10% of 3.3V = 0.33V. That way if the frequency is high enough as not to be able to notice the turning on-off phenomenon with our eyes or ears (imagine a light turning on and off at 1KHz), adjusting the duty cycle actually adjusts the level of something, e.g. how bright an LED shines.

Other waveforms:

There are other waveforms like sinusoidal, triangular, saw tooth shape etc. up to any shape as long as the signal is periodic. Special place have the sinusoidal signals (left) like the voltage of the power grid, but it takes almost a book to speak about those as all signals can actually be composed by sinusoidal

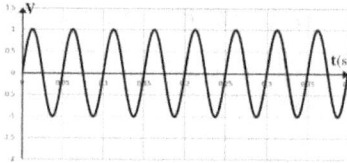

signals. Non periodic signals are very usual also, like the sound signal of a song or a man narrating this book (only the sound signal of a tone is periodic).

Frequency response:

This is a complex matter (involving also numbers mathematically called complex!) for which we will not give an in-depth analysis, just a quick visit since it is in all places in analog electronics design.

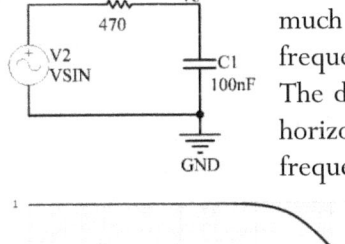

The subject of frequency response is how much a signal is amplified versus its frequency only of **sinusoidal** input signal. The diagram therefor has frequency in its horizontal axis (actually sinusoidal frequency). A magic of sinusoids is that the output will also be a sinusoid all times (it will not distort its shape). As an example let's apply 1V amplitude sinusoidal signal to the circuit on the left by the V2 signal source at various frequencies and measure the amplitude on the "VC" point. In frequency response diagrams you will usually see both axes in logarithmic scale. The vertical axis may be in a confusing unit called decibel (or db). If ever you need decibels calculations check Wikipedia or google on this, we try to keep this subject in its very basics here. What we see in the previous diagram is that in high frequencies signal attenuates, falling to half at 6KHz and after that dropping proportionally to the frequency. This circuit allows frequencies up to 3.4KHz to pass through it attenuating less than 30% (or -3db), we say it has a **bandwidth** of 3.4KHz and is a low-pass filter. There is a bandwidth limit in all electronics, most amplifiers end up at a few MHz, and average cost oscilloscope instruments end up at around 100MHz. You may get used to the frequency response idea by tickling an audio equalizer (e.g. an app in your smartphone of such functionality) to feel what it is to adjust the frequency response of a signal, the audible sound in that case. If we need a square wave signal to pass through such a low-pass filter, its bandwidth (sinusoidal frequencies) should be at least 6x of the highest square signal frequency we need to pass through in order to conserve its rising and falling edges steep enough.

Impedance:

Why did the previous circuit attenuate a signal more the more the frequency it has? Our capacitor charges and discharges at the frequency of the signal. The smallest the signal period is, the smallest is the $\Delta V = \dfrac{I \cdot t}{C}$ across its pins since it does not have enough time to charge or discharge. It behaves, kind of, like a resistor that is smaller the higher the frequency is. That frequency depending resistance-like effect is called impedance, also measured in Ohms. We will see that term in capacitors and coils. In capacitors it is disproportional to the frequency and proportional to the capacity, in coils it is the opposite. All those said in some simplicity since "complex" numbers mathematics are required to make accurate calculations there.

2.6 ACTIVE COMPONENTS AND ICs: REGULATORS AND OTHER USEFUL

In this chapter we will learn more about active components in terms of electrical functionality. This will be the final chapter of components excluding MCUs. Comparing (in fantastic terms) a circuit to a human, MCU is the brain and all the other cells and organs are the components. Only a brain cannot be sustained in life and if so, it cannot do much itself without the rest of the body's organs (components). We will see the very practical ones. Yes, they are a lot more functional and exciting than resistors, capacitors and LEDs.

LINEAR REGULATORS (AND LDOs)

Having said the most about linear regulators in chapter 1.7, it is worthy to repeat some and mention a few more since regulators power almost all electronic circuits. Let's use another one in our tutorial, the MCP1700T-3302E/TT from Microchip Technology. That's in a nice tiny SOT-23 SMD package offering very low current to operate itself (Quiescent current), very low voltage dropout making it definitely a Low Voltage Dropout – LDO regulator and low cost. **When we meet a new component the first thing we should always do is to download and open its datasheet!** As said again, digikey.com is a great fast way to do that or any other distributor's

3-Pin SOT-23

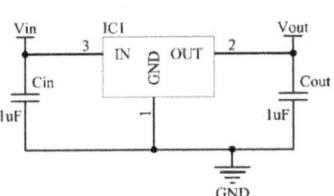

site that suits you. On the circuit on the left we see again the full circuit required, the same in 99% of all linear fixed output regulators. This one outputs 3.3V fixed Vout. The input (Vin) voltage required is from 3.5V to 6V. How much higher the input voltage has to be than the output is the dropout voltage and in this one it is around 200mV only (an LDO). Let's see what to take care about in the process of choosing a linear regulator and the traps that are difficult to spot in a datasheet or the front-end specs.

1. Maximum output current: Here is the biggest trap of all. This specific datasheet for example clams to provide 250mA. Never trust this specification since it is specific to conditions that may not be met in your application. Linear regulators behave like resistors! Adjusting their resistance automatically as to provide the fixed output voltage to their load. That resistance is a controlled "activation"

   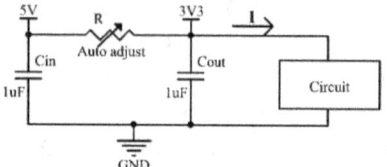

   of an internal power transistor. If in an instant I = 200mA that resistor should be R=V/I = (5-3.3)/0.2 = 8.5Ω. Regardless its value, the thermal power it dissipates is V·I = (Vout - Vin) · I = (5 - 3.3) * 0.2 = 0.34Watts. Will it "take that heat"? To answer that there is a specification called Thermal Resistance (junction to ambient) in °C/W that is how many Celsius the temperature will rise from the environment's temperature per thermal Watt dissipated. In our case it is 212°C/W if it is mounted on a square inch of copper pad on a PCB, if not it may be even the double. Even in that case, 0.34W will make it go 0.34 * 212 = 72°C higher, taking it to 100°C in room temperature. So… for taking care that the maximum current you will ever need is going to be delivered, multiply the maximum (input minus output) voltage to the current and see if the thermal resistance is low enough to give you a go or a no-go. Actually the more output current you need given the input voltage, the bigger body of regulator you need. Bigger is lower thermal resistance. Going to very high current output will require THT packages with aluminum coolers.

2. Dropout voltage: Do not choose for example a 0.9V dropout if your input voltage in the previous example may go as low as 4V.

3. Output capacitor requirement: Some very low dropout regulators counter-intuitively require a special ESR specs capacitor of ESR higher than some value ruling out ceramic capacitors and making a big pain to find your right capacitor.

4. Availability (popularity) and cost.
5. Other specs according to special applications also, like quiescent current in low energy circuits for long lasting battery applications for example.

DC/DC (SWITCHING) REGULATORS

Linear regulators are bound by the following issues:

- Input voltage has to be higher than output voltage
- Input current equals output current
- Power (thermal) dissipation is very high if our input voltage is a lot higher than the output voltage!

But they are simple. Also the output voltage they produce is very stable and has very little noise, call it rippling, usually less than 10mV.

There is another kind of regulators which do magic with coils and

- May output higher voltage than the input voltage
- May draw less input current than the current they deliver in the output
- They have almost no thermal problems in big input to output voltage differences, e.g. may have 12V input, 3.3V output 1A output and dissipate only 0.3Watts. (instead of around 9W)

They are more complex (require more components) need big care in construction or PCB design to have some wires (or tracks) sort and thick, they have about 10 times the output noise of the linear regulators. Most are one of those two kinds: Step-down or step-up. Interestingly there is also the inverter kind that produces negative voltage!

Step-down or "buck" switching regulators:

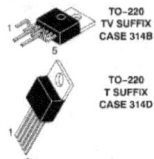

Let's talk about a very popular one, LM2596. It is a 5 pin SMD or THT. As usual, you should start from its datasheet if you ever start to design with it. Here we will only get a first idea of what is involved in such designs and see some specs and benefits of step-down converters. LM2596 is an "old horse" very popular and quit "beefy" in size regulator handling up to 3A in output current. For less output current other very small exist as thermal dissipation requirements are low (thus thermal resistance can be a lot °C/W). Here is a list of design facts:

- In all DC/DC (or switching) topologies, the output (delivered) electrical power ($V_{OUT} \cdot I_{OUT}$) always equals to around the 85% of the input (consumption) power ($I_{IN} \cdot V_{IN}$). That ratio is called efficiency. The rest (e.g. 15%) goes into heat (power in Watts) and that is usually not that much. It is also spread in the IC and in a peripheral diode.
- In step-down topologies, Input current is significantly less than the output current. Example: Taking 12V into 3.3V, 0.5A output current makes about 160mA input current in order to hold output (delivered) power a little less than input power. Drawing less current for operation is a bliss in batteries.
- Adjustable regulators (even linear adjustable regulators) use a voltage divider of 2 resistors to define the output voltage we require. Replacing one of those with a trimmer or potentiometer makes them manually adjusted.
- Dropout is usually more than 1V
- Bill of Materials usually includes the IC, an input and an output capacitor which strictly have to be of very low ESR if ceramics cannot do the job, a coil of specs found in the datasheet, a Schottky diode that can hold the output current and 2 resistors in case of adjustable output ones.

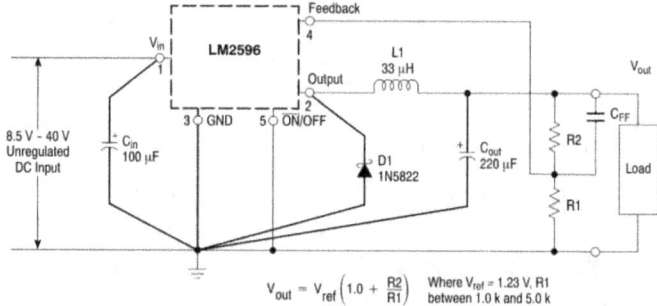

- Some connections have to be sort with thick wires. In PCB designs the ground has to be a copper flooded area. Many implementations on a breadboard will fail!!

- Do not trust the "maximum current" specification. Choose an IC that can provide the double current than the maximum of the current you require. That will also keep it cool.

Considering those, it is not a bad idea to buy small modules – already assembled PCBs that offer just 2 pads for input, 2 pads for output  and a trimmer to set the output voltage. Their cost is really low and your drawer should have some of various specs. It is more expensive to make than to buy if you get those from China (e.g. Aliexpress)

Step-up or "boost" switching regulators:

Same apply. They perform the reverse function. Beware that now the input current is higher than the output current. Useful e.g. to make 5V from a single cell Lithium battery that is 3.3 to 4.2V. Their dropout is usually 0V, meaning that 5V output regulator may work even with 4.99V input. The advice is again to use ready modules when possible. The current specification on them usually applies to the input current that is the higher, in boost it is the input current, not the output current delivered to you.

A final note: Wherever there are switching regulators with electrolytic capacitors, the most common cause of failure is that a low-ESR and low quality cap has got old, its ESR has raised and so requires replacement with a fresh one of as low ESR and as great quality as possible. This is a tip for repairing stuff or giving more money when a system is built to last. Note that the failure has to do with the magnitude of the current value, low currents will not wear them quickly. Also note that caps store voltage and you should not touch such connecting to main supply immediately after power removal.

SWITCHING STUFF OF >20mA

A GPIO has very limited current sourcing capability. When we need to turn on/off many bright LEDs, motors, buzzers or other such current hungry stuff (called "load"), our GPIO can drive or "command" a more powerful switch like:

- A p-channel MOSFET: When the input voltage is the same of that of the MCU supply, the circuit following can provide

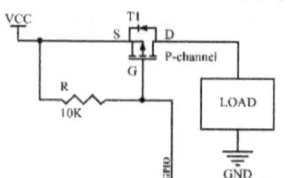

that switching using all the switching capabilities of a p-channel MOSFET. When GPIO is high (outputs Vcc) MOSFET's Gate's voltage (V_{GS} actually) is zero, it is in "off" condition. When GPIO goes low, V_{GS}, that has to be negative in p-channels, is big enough to turn it on. The MOSFET has to be chosen as to handle the maximum current, better with a margin 2x and most importantly to have such V_{GS} threshold (low enough) that it will turn fully "on" when that voltage goes equal to Vcc (3.3V or 5V). 2A driving p-channels cost a few cents. Note that there are "load switch" ICs that can handle higher input voltages, driving correctly for our job an internal p-channel MOSFET. That topology is generaly called "high side" switching.

- Using N-channel or NPN transistors: Tricky but really very useful topology to switch on and off anything. It is important to understand this "low side" or **"open drain"**

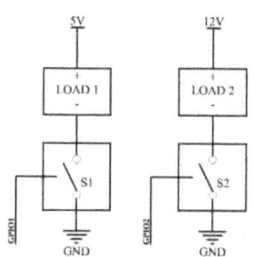

or NPN output switching concept. It is usually the first and best choice too. On the circuit on the left imagine we have such systems (boxes with the switches) **that connect only to the ground** (take that as a limitation and a feature too), they have a "command input" from a GPIO and the other switch's pole open to place our load. Now take a look at the load. Its positive supply has to be connected permanently as it is the

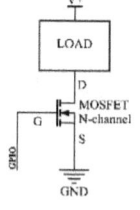

negative supply that connect to our ground or stays floating. That's the limitation that at 90% of the cases is not a problem. The feature is that our load can connect to **any power supply** as long as its voltage is not too high for our switches. We may use a discrete n-channel MOSFET like we have on the left, choosing it as to have low enough V_{GS} threshold to turn fully on with our Vcc that is output by our GPIO on its high state. As you can see MOSFETS provide their drain to connect the load, thus the

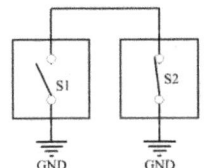

"open drain" naming. MCUs GPIOs also have **open drain output mode**! limiting to their maximum current around 20mA and to the voltage at Vcc. What if two open drain GPIOs (outputs) connect with each other (left) and one is high while the other is low? No problem! We will see that in I²C communication at least. Notice that the high state of an open drain GPIO is floating, the low state is "short to the ground". That requires a pull-up resistor to provide a positive voltage on "high" state.

Very useful in this concept of operation is the **ULN2803** IC.

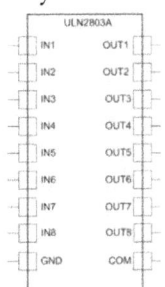

It uses BJT transistors instead of MOSFETs (it is an "old horse") providing 8 open collector outputs (since BJTs have collector in place of drain). It can switch up to 500mA, >30V loads having limitless uses, especially in applications of driving many LEDs such as many 7-segment displays, LED arrays etc., it even goes to driving small stepper motors and any other nice stuff. It is a chip without power supply input.

o Using relays. We described relays in 1.11. When our switching state changes very occasionally and / or the power of voltage switched is high they are the choice, adding the fact that they are real and practically ideal switches. But, in the case we do not want to buy a relay board with its driving circuitry, how do we drive their coil? Simple enough, we use one of the previous methods since a relay's coil needs around 100mA to do its conduct attracting electromagnet job. A note here is that when we drive (switch) coils, there is a problem when we switch them off. Since they react as to keep the current flowing in them as constant as possible, in that event they produce a

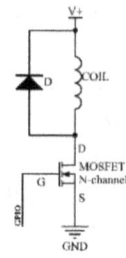

very high voltage themselves (called back-EMF) in their effort to keep current - that now wants to go to zero - steady. That usually destroys our switching component. The solution is to place a diode as in the circuit on the left. That diode is already included in each output of the ULN2803. We have to connect the COM pin to the positive supply though.

Analog signal switches: ICs that act as relays for analog signals of negligible current only (e.g. choose an audio input source). Start from Analog Devices Inc if ever need.

DIGITAL LOGIC ICS

In the dark days when MCUs were not existing yet or were too expensive, many simple ICs did a lot of digital functionality. They still exist, we will visit some still useful in our software dominated era. All are cheap and old school classics.

- o Shift registers: They are parallel input or output. Input type have 8 digital inputs (the parallel side), a pin commands to "grab" or latch their state and then two signals, a clock input and a data output provide our MCU with the 8 parallel inputs states bit by bit, serially (shifted one after the other at every new clock cycle). Output shift registers do the opposite (set their 8 outputs). When GPIOs are not enough in our MCU, shift registers come to the rescue to act as GPIO extra inputs or outputs. They can also be connected in chain, e.g. 3 acting as an 8x3 = 24 port system.
- o Flip-flops: No, not for the beach, a flip-flop latches its state making it a 1 bit memory storage device, useful to toggle something by a button press for example.
- o Timing: Delays, one pulse generators etc. The most classic chip there is the most famous chip of all time (there is a whole book written containing applications made only by it), the 8 pin 555. You can make an LED blinking with it for example. Rarely used today.

o   Logic Gates: If you have read at least one older written book for electronics you should wonder: "why have we not talked about them yet?" Well, if we approach electronics mathematically they are important. In our practical and physical approach we will see them only in software. Myself in the last 20 years I have made use of only one gate. They are the famous AND, OR, XOR and the inverted output NAND, NOR and Exclusive-NOR, doing almost all combinations one can do with 2 bits input (00,01,10,11) and one bit output (0,1). There is also the one input / one output inverter that may be some times handy. An inverter outputs 0 when its input is 1 and outputs 1 when its input is 0. We will speak about gates in a software related chapter later. Note that all our magic computer in our MCU chip is comprised only of gates, some hundred thousand to millions but they are in the micro cosmos of the silicon of the chips and we are not about to design silicon (yet?).

AMPLIFIERS

Strictly on analog signals, we need some times to amplify some very week for making them big enough to be able to measure them or we need to make some powerful enough to drive a big loudspeaker for example. The first category is <30mA consuming ICs with a vast set of specifications especially about noise and bandwidth. In that 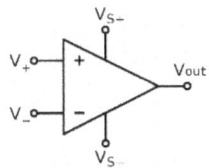 category there are some special "Swiss army knife" style ones called "operational amplifiers" or op-amps (left). There are standard topologies that make them a voltage amplifier, an inverted output voltage amplifier, a voltage to current converter, a voltage comparator, filters of many kinds etc. There are also instrumentation amplifiers that just do great amplification of a signal with very great specs in accuracy. The second category is usually of complete circuit boards, big and heavy according to the power Wattage they can deliver. In those there is a category called "class-D" that works like the DC/DC converters dissipating very little thermal energy per Watt of output, therefor making amplifier boards smaller and more powerful (if power is the main concern).

A POTPOURRI OF OTHER ICS...

For completing the most of the picture, here is a list of others very much useful:

- Battery management: Lithium battery chargers mainly
- Motor drivers: H-bridges and stepper motor controllers as we discussed in 1.11
- USB to UART (serial port) converters: In that the Chinese CH340G is a worth mentioning IC. Ready dongles are a must have.
- Communications level translating ICs: There are serial communications like RS232, RS485, can-bus and others, each using different voltage / current levels to convey information in long wires. An IC special for the application is needed at such cases.
- Memories: Flash or RAM usually communicating with the MCU by SPI protocol in as small as 8 pin chips.
- PWM and timing generators: the most advanced and handy are controlled over I^{2}C
- All digital sensors of course

We shall repeat for last time that in every IC we start from its datasheet. If we are about to use it in our design we read and understand it all in every detail unless we copy a circuit from the internet that surely works and we despise advancing our self's knowledge.

2.7 ANALOG SIGNALS AND MEASURING INSTRUMENTS

If electric nature concepts got you tired this far, we will now on relax with description of measuring instruments and techniques. Briefly first we will add some concepts on signals since signals are what we are about to be measuring.

SIGNALS: AC AND DC COMPONENT

The book you are reading is certainly unorthodox in presenting the basic knowledge about electronics. Having visited the basics regarding voltage, current and signals and gone up this far into intermediate knowledge, we have not yet mentioned what "AC/DC" is and that is intentional. Sometime in the early 70's somewhere in Australia, Malcolm and Angus Young developed the idea for their band's name after their sister, Margaret Young, saw the initials "AC/DC" on a sewing machine. AC⚡DC made it to become a great band and bang our heads a lot of times. Closing this fact, "AC/DC" actually means "alternating current/direct current" in electricity. Direct current is when its direction never changes. Same applies for direct voltage, its polarity always keeps the same direction as in our so far cases, more positive than the ground, always. Alternating current/voltage changes its direction/polarity. A negative voltage signal will be of lower voltage than the ground. Fortunately in practical modern electronics that is a rare case. Negative voltages are not permitted to be applied to any MCU pin. They are needed sometimes in analog signals that may be alternating by their nature, which in turn make us to supply amplifiers and other circuitry with a positive and a negative supply e.g. +5V and -5V as to operate in a range of negative polarity also. Ground is always the zero voltage.

Here is a very important concept that is confusing: the AC component of a signal. Take those two signals, a DC and an AC:

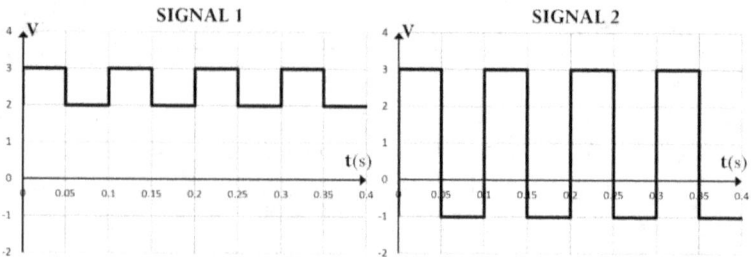

Only the signal (2) is alternating. Even if (1) is not steady, it keeps always the same direction. We call **AC component** the part that fluctuates. In signal (1) it is 1V peak to peak, in (2) it is 4V peak to peak. We call **DC component or DC level** the average value of any signal, in (1) it is 2.5V in (2) it is 1V. So a signal fluctuates around its DC value by the AC component. Since AC component is an amplitude, we have to define if it is peak to peak, just amplitude or another kind we have not talked about, **RMS**. RMS is a usual term we meet in measuring instruments (which is what this chapter is about). It is about the 70% of the amplitude (amplitude is the half of the peak to peak amplitude) of an alternating signal. Actually it is mathematically more complex, it depends on the waveform shape, if our signal fluctuates around zero (DC level = 0) its RMS voltage is how much DC voltage would cause the same heating to a resistor.

Enough about signals, let's go to measuring.

A PRIMER TO MEASURING

Imagine you want to invest a million dollars, you are in a prison inside a scammers section and you want to give your money to someone to invest them for you. That is a gloomy picture of how you should feel every time you measure something! You may be scammed by many physical phenomena, by the instruments you use and by yourself making mistakes in the process. Also there is no way to measure something exactly (except perhaps counting number of events with a reliable method). If a real voltage is 2.66484324851568654845375437 and infinite more digits (we will never know this "absolute truth" of course) we may in best case measure 2.6648432 ± 0.0000005V with a 10,000$ instrument following a very careful procedure mostly regarding the environmental conditions. In 1.11 we said the basics about error, mistake, accuracy and precision. Measuring is the ultimate engineering, never think as a mathematician or a philosopher when you are making your measurements and when you are processing them.

Multimeter

A digital multimeter is the king of instruments, it is the most important instrument, without which we can do almost nothing with electronics. It is a volt-meter an amp-meter and an ohm-meter at least. Fortunately even cheap ones costing just less than 10$ can do most of the jobs and actually amazing well, but since it is our eyes and ears we should spend 40$ at least on that. Some old-school guys may present to you an analog (needle and scale) multi-meter, just laugh at their faces if you see such an outdated in all specs instrument

(not all retro instruments are bad, voltmeters are). We will see how to use a cheap one since they are most complicated as they have no automatic scale functionality. Using an auto-scale afterwards will be straightforward.

Voltage measurement: The two probes have to be placed on the COM and the V input sockets. Current measuring input or measuring mode must not be used for voltage measurement ever, since current measuring is a sort circuit within our instrument. That must be always a concern, even dangerous if trying to measure the mains voltage. With the rotating selection switch we choose a scale on direct $\overline{}$ or alternating $\sim$ voltages. Here is the use of our previous paragraph: DC selection will measure only the DC level of our signal (it may be fluctuating, it may not) and AC selection will measure the AC component **only** of our signal in RMS (regardless its DC level, a battery for example will measure zero). In DC voltage if the value is displayed as negative it means red probe is more negative than the black probe. What are those scales? In VDC for example we see 200mV, 2000mV (=2V), 20V, 200V and 1000V scales. We have to select manually the smallest scale for our signal. If e.g. we tap our probes on 3.45678V, on scale 200m we will see ".1" meaning out of scale (there is never danger to damage our instrument in a more sensitive scale), on scale 2V will still see ".1" on scale 20V we will see 3.45V on scale 200V we will see 3.4V, on scale 1000V we get 3V. You understand that the right scale offers the highest resolution. Not to be bothered with that in auto-scale instruments. All multimeters in volt-meter mode act as an open circuit, not affecting our circuit by measuring it. That is actually about 10MΩ resistance between the probes.

Resistance measurement:

For measuring how much a resistor is we choose again the Ohms scale the same way, the smallest scale possible. That looks straightforward enough, but measuring resistance has three pitfalls: a) Our skin has a resistance of 10KΩ to 100KΩ (we can measure that as well). When we touch <u>both</u> of the probes to keep them in touch with the resistor's leads, we place that (human body's) resistance in parallel to the resistor measured. That is no problem for measuring resistors <100Ω and big problem for >1KΩ, making bigger the deviation (reading lower than real value) the bigger our resistor is. We should touch our resistor only on its one lead with our hand! b) If we measure a resistor that is soldered on a board or connected to other components, our measurement will be very wrong due to other resistances in parallel as well. c) Even worst if there is voltage applied or charged capacitors in the previous case since our instrument measures resistance by applying a small current and measuring voltage.

Exactly the same rules apply to some multimeters measuring capacitance.

There is a special resistance measuring mode called "continuity testing" that uses an internal beeper, usually displayed with this •⁾⁾⁾ symbol. Our instrument beeps when its probes connect together by a wire (or are sort-circuited). This is very useful to see if something connects directly to something else. It actually beeps in any resistance lower than around 30Ω, voltages should not be applied of course. That mode usually carries a symbol of a diode also. When we measure a diode, it measures approximately its forward voltage drop when connected positive probe to anode, negative probe to cathode and out of scale (open circuit) in reverse. That is a quick diode tester. At some LEDs is makes them shine a very little bit also.

Current measurement:

For measuring current we must cut-off a wire and connect in-series to it our instrument that has to act like a wire (zero Ohms). Our probes so are sort-circuited and we have to be extra careful at what we will touch with them. Never also leave a multimeter in current measuring mode when the work is done. There are two inputs (probe's sockets) for current a [mA] input and an [A] input (usually 20A). The latter is for the higher scales. If we open our multimeter we will see inside a thick wire connecting COM and 20A input directly. The 20A (in some 10A) input is an input we trust of being a practically ideal sort circuit so it will not act as a resistance, adding e.g. a few Ohms resistance in our circuit so making the current that flows less than it should be while we are measuring. Note that in high currents e.g. 5A, even 0.1Ohms make trouble (0.5V voltage dropping). That is usually the only problem of cheap multimeters, not the multimeter itself, but the probes cables, having thin wire in them. You should own a pair of good quality thick wires probes. All multimeters have no protection in that current measuring input (a fuse for example). That's not so bad. Actually, momentarily they can hold 40A at least and it is very hard actually to burn one (if probe's wires will not burn first). Also, they will not burn a fuse at any case and their resistance is zero. The only counterpart is the low resolution at one scale only, 20A, providing 0.01A resolution usually. The [mA] input works at the rest of the current measuring scales but has a great problem: In most instruments it is limited to 200mA, exceeding that burns an internal protection fuse that we have to replace (and own some spares) by opening the instrument or its battery compartment at least! They also have a very few Ohms in series resistance, higher the lower measuring scales are, so it is not an ideal current meter. But measuring e.g. 10mA with 1Ω in series resistance makes a voltage drop of 10mV only affecting the current flow very little in most cases.

LOGIC ANALYZER

We will be covering our measuring equipment from most important and affordable to less important/affordable. The next one we should  have in our MCU dominated world is a logic analyzer. What is a logic analyzer? Imagine making a circuit using some GPIOs of an MCU configured as digital inputs (measuring only "1" or "0") and a program recording their values very fast. Displaying then this recording in our PC (over a USB connection) gives us the ability to

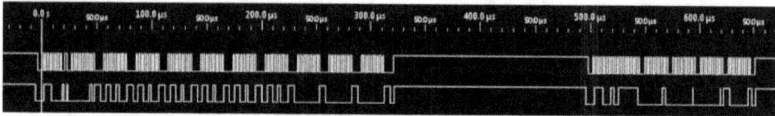

see what information is conveyed by serial communication protocols (UARTs, I^2Cs, SPIs) or how other digital world signals behave so that we can find out how well programs work, mostly in time accuracy or where is a problem in data transmission mostly. A logic analyzer 10 years ago costed over 2000$, now great devices for the job cost 5$!! A PC has to be on our bench, as always. They can sample at speeds over 20MHz (or Mega samples / sec) up to 8 or more inputs. Since we want to see only around a very specific area in those vast data produced, the trigger concept is introduced. Trigger is an event occurring in conditions set by us, e.g. on the first rising edge of an input. Trigger event is usually the time zero, before that event time is negative e.g. -2,-1,0,1,2,3...

OSCILLOSCOPE

Oscilloscope is kind of the instrument "of a wealthy man who has 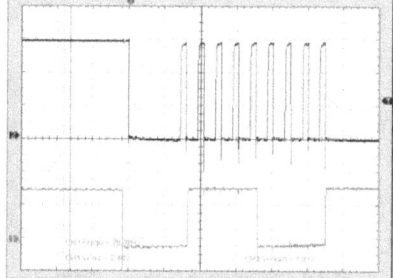 them all". It measures and plots voltage vs time. Having an oscilloscope (a "scope") we see almost anything in our circuits (voltages only). We measure well any timing matter (frequency, period, duty cycle etc.), we see any fluctuation (discovering always unwanted ones like noise, spikes, short timed deeps in supply voltage etc), see the real shape of pulses etc. It is amazing how analytical we can see in time scale, medium grade scopes provide up to nanoseconds per horizontal grid division (time/div) scales. They are specified mostly on how many input channels they provide (two are mostly needed), how fast they sample voltage (samples/sec, anything more than 100MS/sec is awesome), their input signal bandwidth (anything more than 100MHz is awesome, this is for sinusoidal signals, 100MHz is good for square waves up to around 20MHz) and other features. Old times scopes, called analog scopes had a phosphor CRT screen without capability to store or hold steady a measurement (graph). They are awesome for analog signals still but in our digital dominated world we should prefer a Digital Sampling Oscilloscope (called also DSO). They may be separated in two kinds, those with their own screen and buttons and those that use a PC for display and user interaction. Over the same money do not choose a small screen scope versus your full HD monitor, you will lose terribly much information on your signals in a small low-res screen. It is advised to spend over 50$ on a DSO, preferably around 150$ if it is a USB one using your PC and over 200$ if it goes with its own screen and buttons. High-end scopes go up to GS/sec and GHz bandwidth but your signals will very rarely be over 10MHz. Chinese Rigol and Hantek are some nice value for money brands as well as others, while on the high-end we have the amazing Tektronix, Keysight and others.

On a scope we also have DC and AC "input coupling mode" like in the multimeter. AC is useful to see small fluctuations "riding" a steady big voltage (DC value) like fluctuations of mVs on a 5V power supply line for example. DC coupling shows the signal as it is. Trigger is a core functionality when we try to grab some non-periodic events like a once in a while occurring pulse or when we need to hold steady a periodic signal on our screen. You should note that all scopes have their ground connected to the electrical grid's ground, so do many other USB connected devices (connected to a PC connected to the mains) and that must prevent us to connect any ground clip of its probes to any positive or negative voltage.

SIGNAL GENERATOR

They are not measuring instruments but they are very useful to do quick experiments which you will be measuring in our scope. They are devices that produce waveforms of sinusoidal, square, triangle and other shapes (the first two are only important) of selectable frequency and amplitude. Thankfully those days we can get one going up to 5MHz with less than 20$ if you search for "signal generator DDS". Note that this is not a "must-have" device, but very useful if you engage in analog electronics like amplifiers and filters or need some crazy science quick experiments.

LAST AND VERY IMPORTANT FOR OUR BENCH: POWER SUPPLIES

Each circuit needs a power supply. You also need one for testing motors and other devices which just work if voltage is applied. One thing is to get a fixed voltage. Another much greater is to adjust your output voltage smoothly with a knob (potentiometer for example) in order e.g. to find out your minimum operating voltage easily and do many other experiments.

Power supplies act like voltage sources of very little internal resistance. Besides the easily understood spec of their voltage output range, the spec of current output defines the maximum current they can deliver (using less than that is a bliss, using near that max level is not recommended). Good power supplies also can adjust the maximum current they can deliver by providing an **adjustable current limit**. That makes them safer to our circuits in the bad case they behave as sort-circuits. They can stay sort-circuited forever delivering the max current set. They can also serve as battery chargers where the delivered current has to be limited to a set value. That makes them another kind of supply source we have not seen so far, the **current source**. A current source is an auto-adjusting voltage source that lowers its voltage so that the current will be kept at the set limit. If its load (supplied device) resistance goes higher than some value it will not output voltage higher than that set to the supply, so it will output less current up to the point of zero current at open output. For example if we set our supply at

10V output, 1A limit, connecting a 0 Ohm resistor will output very little voltage and 1A of current, connecting 5Ohms will output 5V and thus 1A of current, connecting 20Ohms will output 10V and 10/20=0.5A of current (1A here should require 20V), open circuit will output 10V, 0A. Practically you adjust the voltage then adjust the current limit by sorting the supply's output and then connect it to your circuit.

The power supplies from lower to bigger money go as follows:

- Zero money: Take an old USB cable, cut it near the USB socket that connects to the device, strip its cables and use the black and the red one. It will provide 4.5V to 5.2V up to 500mA, even more if connected to a phone charger.
- Get a wall pack that outputs 15V or more, 2A output if possible (around 5-10) or a laptop power supply and connect its cable-end to cheap step-down modules costing around 1$ each. In all cases your drawer should always contain 3-5 such modules.
- Do the same as previous with a more sophisticated >3A  output current step-down module that has voltage and current display and current limiting (3$ to 10$). Have at least 2 such modules.
- Be a pro. All previous switching DC/DC modules have a noisy output. There is another kind of supplies from old  days that uses a transformer (big, heavy with lot of expensive cooper) and linear regulators for the job. Its ripple is less than 5mV in light loads, compared to 100mV or more in DC/DCs. A 30V 5A such should weigh about 5Kg but it is worth the money if you are well equipped of other stuff (level of lab equipment 3 and 4). Choose one that has no fan to cool itself since the ones with fan are too noisy (I mean too noisy) and actually of cheaper materials due to less aluminum in their cooler (remember, linear regulators heat a lot). Good money for such are about 100-200 for 5A.

Recommended never to buy: Switching mode supplies with color graphics screen, lots of buttons and dials costing less than 200$. Any such devices make your life harder only to adjust two values (voltage and current), your money go to fancy screens and you get all the badly regulated output with a lot of ripple, sometimes as terrible as more than 300mV.

2.8 MICROCONTROLLER ANATOMY: DEEPER EXPLORATION OF THE ROOMS OF THE MAGIC CASTLE

We leave general electronics knowledge behind and move on to a different scenery, that of a computer. We will be wondering inside a world with CPUs, MCUs, Arduino boards and software. Those are far more functional in making our imagination and thoughts about what our project should do to come true (like "hmm… I'd like when that happens to receive this message and to activate that system"). Let's wonder into the computer world, like in the TRON movie, but for real.

In Chapter 1.9 we show a castle. We will make another round in the MCU castle, repeating some things as to put them deeper into our long-term memory.

A quick refresh in the binary system for start: 4 binary digits (bits) hold up to $2^4 = 2 \times 2 \times 2 \times 2 = 16$ numbers (since they include zero, 0…15 that is 0000…1111), 8bits that is a byte, hold up to $2^8 = 256$ (0…255), 16bits hold $2^{16} = 65536$ numbers and 32bits hold $2^{32} = 4\,294\,967\,296$ numbers.

CURRENT STATE OF THE ART IN <5$ MCUS (EARLY 2020)

A micro-controller or MCU is a full computer with peripherals in one IC. The core that executes the program commands is the CPU, which together with the memory forms the computer. The rest are hanging around that, as rightly named, peripherals.

How powerful computer that is? In STMicroelectronics (ST) products we find up to 48MHz clock, around 48 MIPs execution

speed (Mega Instructions Per Second) of 32bit numerical data/commands size (we call it 32-bit bus CPU) 256Kbytes flash program memory (fitting enormous programs of more than 20,000 lines of code), 32Kbytes of RAM memory (for temporary or changing data) and 51 GPIOs on a 64 pin QFP package. That's the STM32F030RC costing around 2$ for 1-20 pieces. On peripherals it offers a great

performance 16-input ADC of 12 bits (4096 steps) resolution that is very fast (1MSamble/sec), 6 UARTs, 2 I²Cs, 2 SPIs and lots of timers. Those peripherals share the 51 GPIO pins. Having this as a reference (a great value / money choice to be noted) we may start to imagine what we can achieve in designing and making boards of  total components cost of one figure. Another worth mentioning MCU is ESP32 from Chinese Espressif costing, as a module with memory and antenna included, around 2.5-4 offering much greater CPU and memory size, less peripherals in GPIOs and ADC though, but powerful Bluetooth and WiFi communication functionality that can implement easily Internet client and servicing functions. Arduno MCUs (ATMega328) unfortunately are old designs of ICs and even though they cost around 1.5$ each, they are far more mediocratic but very great for automations projects. They are still in great use mostly because of the software compatibility. Anyway most of real projects require less than 1000 lines of code, served by the Arduino UNO adequately. Software in all history of computers is what makes users what computer to choose. To my view, that is right.

CPU CLOCK SOURCE

In a computer's CPU a clock is a square periodic waveform signal at an x frequency. At every period of that clock signal (every clock "tick") the thousands transistors in the CPU make their next operation - that is executing the next command of a program - (exception is in big CPUs of PCs and smartphones which are pipelined, executing more than one command in one clock pulse). Circuits that produce a periodic signal (at a specified frequency) are called oscillators. There are two usual ways to produce the clock signal (with the frequency we like it to have, up to the maximum allowed) for an MCU. Either with an internal oscillator or another internal oscillator that uses an external crystal. The first requires no components but offers frequency accuracy around 1%. Any time-measurements or time-calculations in our program will also be of  the same accuracy. External crystal oscillators require a component called "Crystal" and 2 small value capacitors. The frequency accuracy goes to about 30PPM (parts per million) = 0.003%! Each crystal produces one frequency, so does the internal oscillator, internal frequency dividers and multipliers (PLLs) can convert it to our desired clock frequency.

TIMERS

Since we talked about clocks, the timer peripherals (there are more than one independent timers) use the clock frequency to count time or produce timed signals of their own, not requiring program commands executing in the CPU to do that functionality. The main benefit is that they will never miss one clock's cycle regardless of what our program is doing, functioning un-interrupted in the background. In the 16MHz clock of Arduino UNO, a timer counts up to the number 16000 and interrupts our program to increase a milliseconds counting variable by one. That way we now the time in milliseconds passed since MCU started to execute our program using the millis() function. Other applications of timers are to produce accurately timed PWM signals, count pulses of GPIOs never missing any whatever our program is doing, feeding peripherals like UARTs with a specified frequency made by dividing the CPU clock to a number (count up to 100 for example for making a new pulse should divide our clock frequency by 100) and other such functions. There is a special timer in some MCUs called RTC - Real Time Clock. It provides calendar and hours-minutes counting. Moreover it can use a separate crystal of 32,768Hz frequency (that is 2^{15} for a good reason) and a separate power supply source in order to keep working with a miniature battery while our MCU is off. In Arduinos that functionality is well served by external boards which carry their own coin cell battery, communicating with our MCU over SPI.

ADC

As said in 1.9, ADC means Analog to Digital Converter. An ADC is one peripheral with an analog switch at its input, connecting to more than one pin of our MCU. We select which pin to connect to (one pin) and then we instruct it to measure. Its main specs are the resolution in bits, 12bits is $2^{12} = 4096$ steps (0...4095) for example.

That will convert a voltage of 0V into a digital value (a number) of 0, and the maximum voltage allowed that is usually the Vcc e.g. 3.3V into the maximum of that number, e.g. 4095 for 12bit. All the rest are proportional to that, e.g. 0.33V (10%) will make the 409 number. ADCs do not convert or measure instantly. Fast ADCs found in MCUs take around 1usec to measure or can measure up to 1million times per second maximum. That is 1Msamples/sec since we call one measurement, one sample. All measurements done solely with our MCU have therefor accuracy, resolution and speed limited by our ADC's performance. Note on that the fact that all ADCs have some input noise, if a steady voltage of 0.33V for example is measured, ADC's samples may be: 409, 411, 409, 406, 408, 409, 410, 407,... fluctuating randomly ±2 to ±4 units (counts) around the correct number. If we collect some samples and take their average we fix that a lot. Checking the absolute maximum ratings in our MCUs datasheet, we see that all MCUs pins, including the analog input pins can never accept voltages higher than the Vcc or lower than 0V (reverse). Keep that always in mind. Measuring higher voltages takes a simple voltage divider, measuring reverse (negative) voltages is harder, needing special amplifiers.

Finally there is the opposite of the ADC, the DAC that is the Digital to Analog converter, not found in all MCUs though. It is needed rarely since PWM does that function even more practically in most times.

UARTs

UART communication or serial com will be in your everyday life. Since it is a very simple and very old protocol, it requires us to set a data speed at both sides, if that does not match no data pass through. It also requires to set how many bits we have per frame (7 or 8), the number of stop bits that makes a frame (1, 1.5, 2) and parity check if any (None, even, odd). In 99% of the cases those are "8N1" meaning, 8bits, No parity, 1 stop bit. It is suffice now not to analyze those more. Speed or data rate, called also "baud rate" is measured in bits per second (bps). Those are not any number, the usual ones are 1200, 2400, 4800, 19200, 38400, 57600, and 115200 bps. There are also some higher, 115200 is very handy and common.

Usually the two parties that communicate are our MCU and our PC, we have to set our serial terminal or whatever other software is used for communicating over our serial port to the baud rate our MCU is using. Have in your drawer more than 2 spare USB to UART dongles at all times. Depending on the PC operating system, they appear as "COM" or "tty" devices when properly installed. We also have to choose the proper serial device to connect with when more than one are present. Arduino UNO and others include a USB to serial chip on their board.

UART has an RX (receive) and a TX (transmit) pin. Two UARTs (MCU to MCU or MCU to USB2UART dongle) have to connect one's RX to the other's TX at both 2 signals. Take some care not to connect TX to TX since you connect two outputs together. Also take care not to connect a dongle set to 5V (5V or 3V3 is a selection set by a jumper) to a 3.3V MCU or vice versa. We will visit the 5V to 3.3V compatibility issue later.

Last, UART is a very easy protocol. Whatever comes to the RX pin of the MCU is well received, byte by byte while we may concurrently transmit anything byte by byte on the TX pin (A bi-directional or full-duplex means of communication). On the data our program is handling, there is no protocol at all. We will visit a protocol next in I^{2}C. We should take the chance here to note the old RS232 interface. That is a UART on strange voltage levels for 1 and 0. 1 is around -10V and 0 is around 10V! RS232 performance is a lot worse than just 0-3.3V signals in long cables at high speeds and is only a pain to use it. In the cases we need to communicate to RS232 devices, ICs like the MAX232 covert 0V-Vcc voltage to RS232 levels. You may encounter the old name "TTL level" referring to the direct MCU pins' 0V-Vcc signals.

I^{2}C

Two wires are needed, one is the clock (SCL) and the other is the data (SDA). It is far more complicated than UART but allows one device called master to connect at the same time to more than one "slaves" and talk with them all. UART cannot do that, so one I^{2}C can be as many as 100 or more UARTs. It goes by two speeds 200Kbps or 400Kbps, no needed to be set since the clock signal provides all the timing. SDA though has to handle both directions of communication (transmit and receive). There is a data protocol well defined that all I²C devices follow. Slaves (sensors ICs etc) have a fixed address, master askes to receive some data from x address slave device or to transmit some data to it. The start and the end of data transmission are special CLK and SDA signal levels combinations. There is also an "acknowledgement" response from a slave, toggling SDA in a special timing. That all makes software a little more complex but it is the de-facto interface for most sensors, special

function ICs and small size screens. Next we see an example of 4 bytes transmission, first group of pulses is the address request, atop is the clock signal, below is the data signal. Both SCL and SDA work in output open-drain mode, each needs an external pull-up resistor of 1K to 10K (higher speed requires less resistance). You should not use wires longer than 1 meter for I²C communication, it is designed to work only "inside the box".

SPI

SPI for Serial Peripheral Interface or "Spy" in nickname, is an ultra-fast and ultra-simple serial protocol with a clock signal. Its speed can be any from almost zero to 20Mbps or more, it is bi-directional (transmits and receives the same time using separate lines) but uses 3 lines and needs one more for every device it connects to. It is simple to describe. There is a master (MCU) and slaves (ICs), master outputs the clock signal (CLK), receives by the MISO signal (Master In, Slave Out) and transmits from the MOSI (you guess…) signal one bit every rising or falling edge of the CLK. There is one "device select" line (pin) for every slave output from the master, only the one selected should respond, usually "active low" i.e. the one that is zero value is active. Like I²C it is not to be carried by wires more than about 1 meter away. High speed and parasitic capacity of wires impose a limitation on this.

Analog Comparators

Each compares two voltage inputs acting "analogly", if one is higher the result is "1" otherwise the result is "0". With an ADC we usually can do that as well.

Watchdog

This is a dog watching out your MCU. If you keep it active and do not feed it for a while it eats your program by doing a reset. Really, let's see what it is... It is a timer that counts down like a timer of a terrorist bomb (no there are no red digits and colored wires), when it reaches zero, boom, it resets all the MCU. Your program should re-start it periodically, if it fails to restart it within the duration of its full count-down probably your program is stack somewhere so your system will remain stuck forever. But the watchdog will save the day by making that automatic reset. In most cases we de-activate the watchdog at the beginning since it is a lousy thing. In Arduino and in most MCUs' software it is by default de-activated.

Programming your MCU's Flash Memory

The program made on your PC is "downloaded" to you MCU by either a special interface for that (SPI, JTAG, SWD, the two later are for that sole purpose) or by a UART and the use of a software that has to be already in your MCU called "bootloader".

The Bootloader has to "listen" to the UART when the system starts-up and if data denoting in a pre-defined protocol "we have a program for you" arrives, it starts to take the program data byte by byte and write them in the flash memory. Note that flash memory can be written about 10,000 times in its whole lifetime and that writing any byte of it requires to make a procedure of erasing and re-writing a whole group of bytes (a page that is 256 bytes or more). So flash is considered a permanent memory that may hold in very rare cases new data written to it by our software.

Bootloader is simple but needs the bootloading program somehow to be written already ("flashed") in the flash memory. Note that Arduinos are using this method, there is a bootloader already written in the MCU that welcomes any new program by the UART port. Bootloader stays there forever. If you purchase a new MCU chip there is no Arduino bootloader in there!

ATMegas can be programmed using a programmer dongle that connects to them over SPI bus. Fortunately nowadays it is very cheap (<3$), it is good to have one, though you may never need it.

Other MCUs, especially "ARM core" MCUs use JTAG (about 4 wires) or SWD (2 wires) programming interfaces with special USB dongles.

2.9 THE HARDWARE OF THE ARDUINO UNO BOARD AND OTHERS EVEN GREATER

As for every IC we start and cover all the way to knowing about it reading its datasheet, we will do the same with Arduino UNO's MCU, ATmega328P, staring from page 1 and 2: (note the datasheet is for a series of ATmegas)

 **ATmega48A/PA/88A/PA/168A/PA/328/P**

megaAVR® Data Sheet

Introduction

The ATmega48A/PA/88A/PA/168A/PA/328/P is a low power, CMOS 8-bit microcontrollers based on the AVR® enhanced RISC architecture. By executing instructions in a single clock cycle, the devices achieve CPU throughput approaching one million instructions per second (MIPS) per megahertz, allowing the system designer to optimize power consumption versus processing speed.

Features

- High Performance, Low Power AVR® 8-Bit Microcontroller Family
- Advanced RISC Architecture
    - 131 Powerful Instructions – Most Single Clock Cycle Execution
    - 32 x 8 General Purpose Working Registers
    - Fully Static Operation
    - Up to 20 MIPS Throughput at 20MHz
    - On-chip 2-cycle Multiplier
- High Endurance Non-volatile Memory Segments
    - 4/8/16/32KBytes of In-System Self-Programmable Flash program memory
    - 256/512/512/1KBytes EEPROM
    - 512/1K/1K/2KBytes Internal SRAM
    - Write/Erase Cycles: 10,000 Flash/100,000 EEPROM
    - Data retention: 20 years at 85°C/100 years at 25°C[1]
    - Optional Boot Code Section with Independent Lock Bits
        - In-System Programming by On-chip Boot Program
        - True Read-While-Write Operation
    - Programming Lock for Software Security
- QTouch® library support
    - Capacitive touch buttons, sliders and wheels
    - QTouch and QMatrix™ acquisition
    - Up to 64 sense channels
- Peripheral Features
    - Two 8-bit Timer/Counters with Separate Prescaler and Compare Mode
    - One 16-bit Timer/Counter with Separate Prescaler, Compare Mode, and Capture Mode

ATmega48A/PA/88A/PA/168A/PA/328/P

- Real Time Counter with Separate Oscillator
- Six PWM Channels
- 8-channel 10-bit ADC in TQFP and QFN/MLF package
    - Temperature Measurement
- 6-channel 10-bit ADC in PDIP Package
    - Temperature Measurement
- Programmable Serial USART
- Master/Slave SPI Serial Interface
- Byte-oriented 2-wire Serial Interface (Philips I^{2}C compatible)
- Programmable Watchdog Timer with Separate On-chip Oscillator
- On-chip Analog Comparator
- Interrupt and Wake-up on Pin Change

• Special Microcontroller Features
- Power-on Reset and Programmable Brown-out Detection
- Internal Calibrated Oscillator
- External and Internal Interrupt Sources
- Six Sleep Modes: Idle, ADC Noise Reduction, Power-save, Power-down, Standby, and Extended Standby

• I/O and Packages
- 23 Programmable I/O Lines
- 28-pin PDIP, 32-lead TQFP, 28-pad QFN/MLF and 32-pad QFN/MLF

• Operating Voltage:
- 1.8 - 5.5V

• Temperature Range:
- -40°C to 85°C

• Speed Grade:
- 0 - 4MHz@1.8 - 5.5V, 0 - 10MHz@2.7 - 5.5.V, 0 - 20MHz @ 4.5 - 5.5V

• Power Consumption at 1MHz, 1.8V, 25°C
- Active Mode: 0.2mA
- Power-down Mode: 0.1µA
    • Power-save Mode: 0.75µA (Including 32kHz RTC)

That's an overview to know what capabilities we have. Note that the ATmega series is now around 15 years old!!! (@2020) The first Arduino was released on 2005 with a very similar ATmega8 MCU the first Arduino with ATmega328 was released in 2008. They all have been manufactured by Atmel, which since 2016 is acquired by Microchip Technology Inc. Do not feel discouraged by that fact, you can make circuits on breadboards by those really old MCUs since

they are also in 28 pin DIP THT package! Another good thing is that we start with a simpler MCU, ramping-up to the knowledge, not jumping to the 5x more complex in peripherals arsenal of nowadays MCUs on our first step.

At some of those specs you may scratch your head about what they might be, no need to learn everything from now, you can google a lot as you proceed to what will be useful to learn for your specific project needs. You also learn a lot more about the hardware the more you learn about software.

Arduinos are nothing more than ATmega "development kits", or boards having all the usually required components for generic use, plus connectors for easy experiments, especially by using DuPont cables. We will start and keep going with the most classic and the most "Arduino" of Arduinos, Arduino UNO *(left photo taken from arduino.cc web site)*

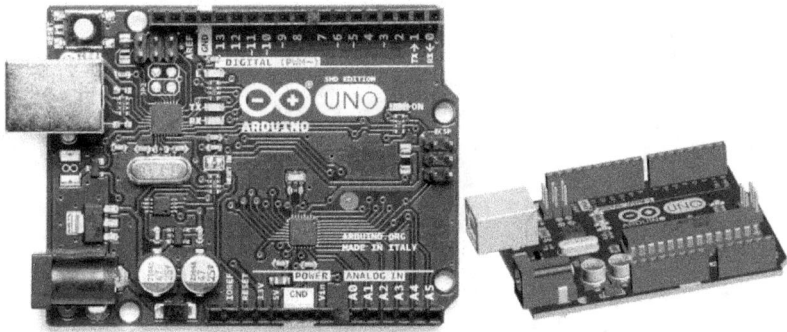

Take a moment to read all those silkscreen markings on the PCB for what each female header pin is. Taking them from left bottom anti-clockwise they are: A group of supply pins, a group of analog input (ADC input) pins that also function as GPIOs if set so, UART's RX and TX, GPIO pins (the ~ ones provide PWM output capability) and finally the two last with no label on the top layer show on bottom layer SDA and SCL of the I²C bus. They are internally connected to the A4 and A5 pins (I²C shares GPIO pins such as ADC does, each pin connects to the peripheral we set it up to connect to).

All there is to know about its hardware is reading all about the MCU from its datasheet and studying its schematic, showing what connects to what. Understanding the schematic also may need reading more datasheets and maybe gaining some generic knowledge, but let's start from the deep. Here is our first big schematic, that of Arduino UNO (revision 3).

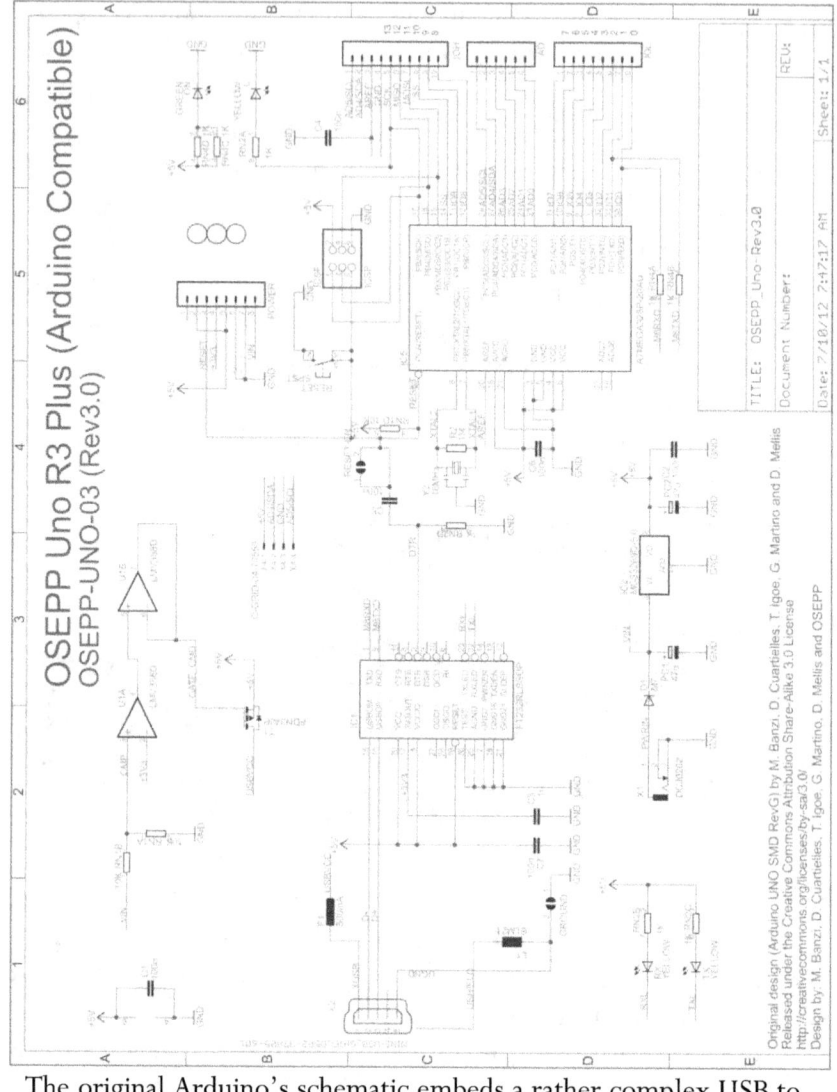

The original Arduino's schematic embeds a rather complex USB to UART circuitry, so I felt it was a bit complicated for looking at it for start. The above schematic is not of the shown PCB but does the same function to whatever is connected to any socket in a simpler way, so it will serve us well here. There are some variations of the Arduino Uno boards on the market. Other are compatibles using alternative but with the exact same functionality components, others are exactly the same with the original Arduino Uno. Since it is an open hardware project they are not "illegal China copies" (neither if you make yours will be any illegal copies) unless they carry the braded name "Arduino UNO" without the word "compatible". Not dealing with the USB to UART bridge chip FT232RL (that needs a datasheet reading for its own), you should understand the most of it. Notice net labels to describe what connects to what besides just drawing wires. That is an easy way to draw schematics, I prefer more wires than labels though since you follow them with your eyes easier. The op-amp on the left works as a voltage comparator to cut-out the USB supply input when the power input socket has a connected supply offering to our board more than 6.6V+a diode voltage drop = around 7.4V. The second op-amp was just there in the dual op-amp chip chosen and actually does nothing, it is configured as an amplifier with gain 1x (output equals input). Be prepared to question some details in boards' designs you will see in the Arduino's world!

5V V$_{CC}$ VS 3.3V V$_{CC}$ CIRCUITRY

In the old days, speaking of middle 90s and back, most digital circuits used 5V as a de-facto supply voltage. Arduino design caught-up the digital world of early 2000, taking the nowadays awful decision to use 5V supply (but probably a good choice for then). Almost all MCUs and digital ICs now are supplied with around 2.5 minimum to around 3.6V for maximum, working best with 3.3V or 3.0V. 3.3V is the most usual supply voltage. As years pass, a few are emerging at 1.8V supply voltage. Low supply voltage is lower power (P=V·I) consumption that is gold in the battery operating devices including smartphones.

Connecting digital lines of circuitry supplied at 5V (Arduinos with ATmega MCUs) to digital lines of circuits supplied at 3.3V is really problematic. 3.3V outputs will "drive" well 5V inputs (3.3V will be understood well as "1") but 5V outputs may burn 3.3V inputs 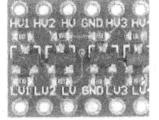 exceeding their maximum allowed voltage. Some MCUs GPIOs (not all) are 5V tolerant while supplied at 3.3V. In most cases we need special circuits (you will find easily) called **logic level converters**. Some cases are resolved by using just a voltage divider (5V to 3.3V)

BEYOND ARDUINO UNO

Others of greater performance than the Arduino UNO exist. Almost all also at 3.3V power supply. Prepare to find a lot, I mean more than 20 kinds and more than 100 models. Outstanding are:

- o 100% like Arduino UNO with ATmega MCUs, more or less pins and memory: Arduino MEGA 2560 provides a lot more resources, Arduino Nano is like a big DIP chip that thanks to its male pin headers snaps well on breadboards.

- o ESP8266 based boards with most notable the "NodeMCU":  Lilliputian on GPIOs and peripherals, carrying the great ESP8266 MCU with WiFi radio capable to do awesome things over internet or your local WiFi network, costing around 2$! 3.3V supply
- o ESP32 based boards with most notable the ESP32  development kit: carrying the mighty dual core ESP32 that does what ESP8266 does plus Bluetooth 4 plus lots of memory (4MB flash, 520KBs RAM!!) and around 20 GPIOs, costing around 4$. 3.3V supply
- o STM32 MCU arduino compatible boards like the Blue Pill  STM32F103C8 and others: Offering the Arduino Nano style with ATmega328 x 10 capabilities and an awesome ADC, costing even less than the Nano. Need to read extra instructions for about 2-10 minutes, and google a little bit about the fake boards in the market. 3.3V supply.

All those are supported by the Arduino software libraries (or framework). There is a whole universe of great MCUs not supported by the Arduino software. They are a lot more complex to begin with, so we will not talk a lot about those in this starting-up book. It is recommended to get involved with those right after some Arduino framework coding. Note as a good choice (subjectively) the STM32 MCUs with ST's software libraries and IDE tools. This far you will be capable to estimate any MCU's capabilities by reading its first page of the datasheet or viewing its features on distributors (like digikey and others) search results lists. We try to learn how to fish rather than getting cooked fish dishes from this textbook.

2.10 SOME REAL ARDUINO CIRCUITS WITH SENSORS AND DISPLAYS

Let's get into real and well working designs.

PROJECT #1: BASIC I/O (IN/OUT)

Let's begin with a simple concept, a circuit controlling fully some LEDs (D1,D2,D3), even how brightly they shine with PWM control and two inputs for the user (you) to interact (play) with it, a button and a potentiometer. This example is not any useful as a system, it is about using some GPIOs and an ADC input for real.

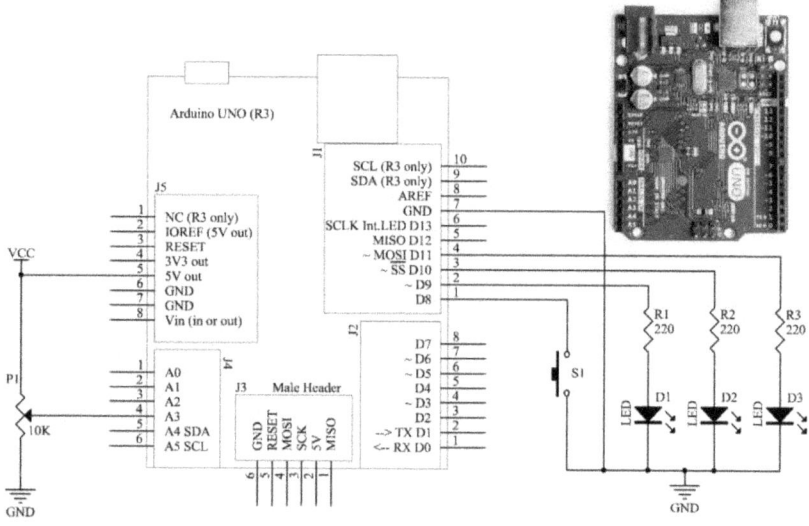

Notice we prefer using as less net names (labels) as possible and use as many lines as possible, except for the supplies and ground nets (net and circuit node are the same, as we have said, all points of a node are connected together therefor have the same voltage). Notice that our ground connects to at least one of the ground pins of the Arduino and our Vcc is connected to the 5V power supply of the Arduino. The Arduino and our circuit with it, should be powered by either the USB socket or by its DC "power in" socket next to the USB socket (7V to 12V DC).

In this project we will take the opportunity to cover various Arduino UNO knowledge that is needed every day (if we get so involved as to have every day an Arduino UNO in our hands of course). We will also pay a quick visit to software commands and settings involved, as to become better friends with it. All software commands are covered in full detail (something like a datasheet) in the Arduino reference page of arduino.cc web site (just google "arduino reference" without the quotation marks). This chapter will be 90% hardware focused.

GPIOs:

Output mode: We need such to be the pins 9,10 and 11. In the setup() section (function) of our program that runs once at startup, we need to place the command (function call): `pinMode(9,OUTPUT);` and accordingly for for 10 and 11. Note that in other MCUs we may choose the output type to be push-pull or open-drain. In ATmega case it is push-pull only, meaning it has two MOSFETs, one connects it to Vcc whem set to 1 and another connects it to GND when set to 0. Open-drain as we said earlier has only the GND connecting MOSFET.

We set the value of an output pin at any point in our program using the command (function) digitalWrite(), setting pin 9 to high goes as: `digitalWrite(9,HIGH);`

Input mode: There are two input modes, input, that is a floating state (connected with nothing) just measuring its voltage and input with a pull-up resistor that uses an internal resistor around 50KΩ connected to the Vcc. That resistor keeps it at 5V (Vcc) when connected to nothing. In our case for the button we have to use the pull-up resistor otherwise while our button in not pressed (is an open switch) pins 8's voltage should be randomly fluctuating. Yes, it is the pinMode() function again, now as `pinMode(8, INPUT_PULLUP);` (non pull-up is declared with "`INPUT`").

We ask to get the value (0 or 1 or equally HIGH or LOW) of a GPIO input at any point in our program using the function `digitalRead(8);` for pin 8 in our example.

PWM output:

So far, by using pins 9,10 and 11 as digital outputs we can either turn on or turn off an LED (make it blink or whatever on-off sequence we can do using time and logic). The 6 pins with the ~

symbol have PWM capability. Adjusting PWM duty cycle (0% to 100%) we can adjust the ratio of time each is on to the time it is off. The frequency is 490Hz at some of those pins and 976Hz at some others, fast enough our eyes not to catch the flickering of the LED's light, so its average value will look to our eyes a continuous, steady brightness (equal to the duty cycle ratio x the "always on" brightness).

ATmega328's timers offer 8 bits resolution to the duty cycle setting, that is $2^8 = 256$ steps (value range is 0...255). So, driving it with PWM of 50% duty cycle will require to use 255/2 parameter = 128. All other ratios are just proportional (value goes 2.55 for every 1% of duty cycle, only integers of course). In our software it is terribly simple, we must already have set it up as OUTPUT and at any point of our program use the command (function) `analogWrite(9,128);` for setting e.g. LED D1 shine at the half of its brightness

Measuring with the ADC:

ATmega328 has a 10bit resolution ADC. 10 bits is $2^{10}=1024$ steps of resolution or "counts" (a kind of poor performance considering some added noise). In our circuit, the potentiometer's wiper (or cursor or middle pin), acting as a voltage divider output will have a voltage of 0V when it is at its one end, Vcc (5V) at the other end and all the voltages in-between at the other intermediate positions.

ADCs have the concept of the "reference voltage" that is the voltage at which reading (ADC measuring result) is at the maximum count (1023). We can select it, the default is Vcc (5V). Therefor it is: 0V → 0 counts, 5V → 1023 counts. So every count in the number our software reads is 5V/1023 = 4.888mV. If e.g. we read 100 the voltage is 100*4.888mV = 488.8mV. Analogies are the 95% of the math you will ever need in electronics engineering. In our case a linear potentiometer will output 2.5V at its middle, no need to do the calculations it will be 1023/2 = 512 counts reading in our software.

Using the ADC is simple in the Arduino software. Having done no setup at our pin (leaving it at its default mode that is input) at any point in our program we use the command (function) `analogRead(A3);` (parameter is A0, A1..A5)

So we can write a software project that will read at any time the potentiometer's position (primary value: 0-1023) and the button state (0 or 1). It can then set any LED to any brightness doing light sequences or anything limited by the imagination (and limited by 3 LEDs). Patience, next chapters will all be about software.

PROJECT #2: AN ENVIRONMENTAL CONDITIONS CONTROLLER

The project's mission: to measure environmental conditions and adjust temperature by automatically turning on and off something like a heater (not included in our project) as to keep the (measured) temperature as near to a set value as possible. This can be expanded easily to controlling humidity or… you name it… We already included 2 things to control (e.g. a heater and a humidifier). Among the great variety of sensors we will choose two intentionally for understanding better the two most common ways of connecting them. ADC input and I^{2}C. For that mission a simple photo resistor will be connected to an ADC input and an advanced temperature, humidity and pressure sensor will be connected to the I^{2}C bus. For user interaction or interface (UI) we will use a very simple screen and just two buttons. The purpose is to present more schematic concepts rather than a super useful and amazing project. Amazing projects will come out of your head plus there are many in the internet also. Learning to fish is better than one big dish of tasty fish. Into our circuit…

2.10 Some real Arduino circuits with sensors and displays   p 172

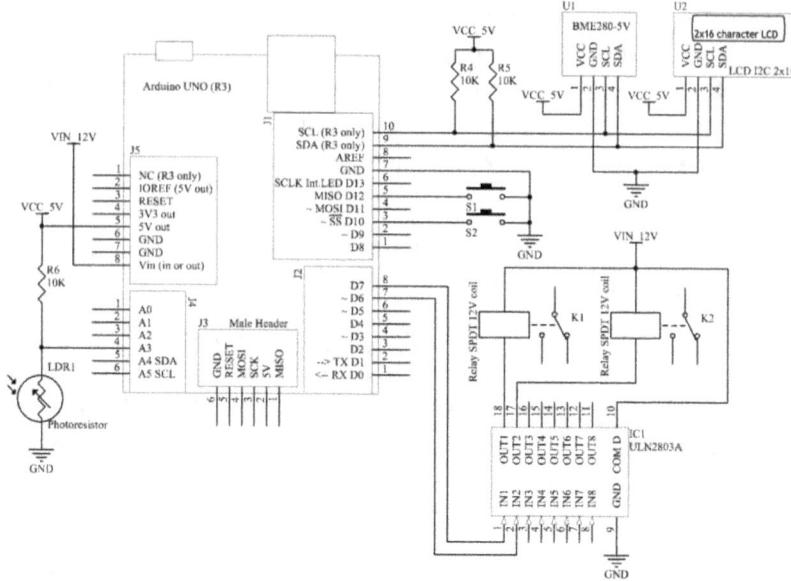

It should be powered by the Arduino's DC socket with a 12VDC supply. That is intentional for driving our relays coils and for understanding better the NPN or open drain switching capabilities. Taken from left to the right:

LDR1 is a photoresistor (we mentioned in 1.11) forming a voltage divider together with R6. In our project it is intended to detect if there is day or night and if the day is sunny. The more the light on it the more the LDR resistance drops so makes the less the voltage on A3 ADC input that the Arduino is going to be measuring.

IC1 that is the ULN2803 we mentioned in 2.3. Each of its outputs connects to the Ground or stays floating according to the logic state on its input, handling up to 500mA of maximum current and up to about 30V of voltage, enough to drive 12V relays coils (12V coils need around 30mA-100mA each if you look into datasheets). We used only the 1/4th of it. Alternatively we might have used relay boards driven directly by the GPIOs. COMD pin is the common cathode of 8 diodes each connected to each output protecting them from back-EMF of the coils. That's the charm of using low-side or NPN or N-Channel transistors: They easily switch on-off any level of a voltage as you can see.

Up and right we have a BME280 sensor board. It is an I²C sensor IC measuring pressure (with extreme resolution), temperature and humidity. We use a 5V board that is "5V compatible" specifically for the Arduino's 5V level of signals.

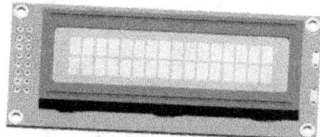
Next to it is an I²C LCD display, the simplest perhaps, a character type of 2 lines 16 characters long each. Those display ASCII symbols. They are of parallel (4/8 data bits + control signals) or I²C interface. The second kind is chosen with the PCF8574 chip for doing the I2C to parallel bus connectivity (or interfacing). Take a glance in the circuit on how two I²C devices are connected. Of course many more can be attached on the 2-wire I²C bus provided each has a different I²C address (the address is a 7-bit number each one has internally for responding or not to an I²C transaction). The two pull-up resistors (R4, R5) are needed at every I²C connection since all the pins connecting to the SDA or SCL are in open-drain configuration. Some I²C boards contain those so they are not required by us (or even in this case they are connected in parallel making no harm). In higher I²C speeds (e.g. 400Kbps) we need smaller resistors (around 2KΩ usually) for faster charging and discharging parasitic capacitances that form at many places on each of the two I²C wires.

Two circuit making examples:

If schematic is understood, let's go a step further, into using this circuit as an example to see two implementation (making) techniques. At the beginning of chapter 2.4 we mentioned:

From the quickest and "dirtiest" to the more difficult and better performing the most common and not only ways to go are:

1. Connect header to header with DuPont cables
2. Use breadboards (and cables like DuPont)
3. Solder THT components wire to wire, all hanging "on the air"
4. Use a prototyping PCB to solder THD and some SMD components, connecting them directly as well as with wires and cables
5. Design a PCB that implements the connections of our circuit, order it and solder our components on it.

Let's start from #1 (DuPont cables). Unfortunately here the connections are quite a lot, so we will need the help of #2 (breadboard) as well. As we have already described that technique, it is a spaghetti recipe. Here is what we cooked:

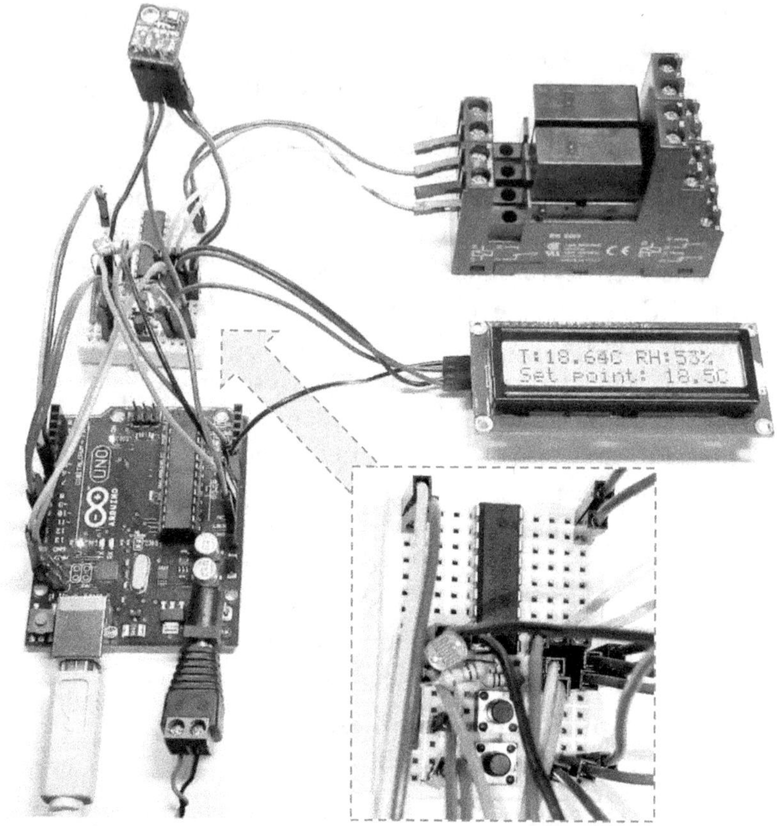

We used a small breadboard here that just fitted, so components and wires on it had been a little dense. The circuit worked well, the advantage of this method is that we can make that spaghetti circuit in less than 20 minutes, so it is a nice way to check our schematic design (not in all cases though, like DC/DC regulators, power circuits, sensitive signals circuits and SMD components).

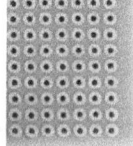 Skipping #3 (solder all hanging on the air) from the previous list, going to #4 (prototyping PCB), grabbing a perforated PCB (pad board) with individual pads (left), our components, some thin wire, solder

and a ton of patience and mood for making things, after a couple of hours we get to a circuit like this one:

Which is looking great on the top side of the picture, is spaghetti on the bottom side, but all are soldered and will keep working for ages even if we move it, shake it or drop it. Its size is also a lot smaller. In this implementation we made a change as you can see on the Arduino we used! Arduino NANO is 98% similar to the UNO but is way more breadboard and prototype-PCB friendly with its male bottom side pins. (NANO has the same MCU, the same pins except a couple which are almost never needed and lacks the DC power "jack" socket).

If those are well and deeply understood, you are quite a good designer of basic electronics already. ☺

Let's move on to the world of information…

2.11 ALL RIGHT, IT'S QUITE EMBARRASSING TO ASK: WHAT IS REALLY A COMPUTER?

The more the technology makes its magic, the more computers we have in our everyday life but the less we feel their presence! In our MCUs world (embedded electronics) we meet computers inside a wireless mouse, inside a TV remote control, inside a printer and about 50 computers inside our car! (Average city car). Computers with a screen display also have transformed a lot, from a desktop device to a device in our pocket and on our wrist.

In computer programming that is what we will be talking about from now on, we should meet a little more the machine (the computer) we are about to be programming. That chance is given better to us since we are using small computers in our MCUs, easier to understand since we are actually a lot "closer to the machine" – the computer, than we are when working with our PC or smartphone.

A **computer** is a CPU (Central Processing Unit) and memory. It uses computer peripherals to interact with the world, but it is not the peripherals (e.g. the keyboard I am typing at the moment).

Memory stores bits in groups of 8bits (a byte) to 64bits (words) or even more in modern graphics cards and PC's (or MACs) motherboards. Memories have an address bus for selecting the byte or word, a few pins to control e.g. if the selected one is to be read or written and a data bus of so many lines as the word's number of bits. There are some types of memories, RAM or SRAM, DRAM, flash, EEPROM, ROM, internal cache SRAM etc. Internally they all work using an address bus, a data bus and a few control signals.

CPUs are the brain, but let's see how smart they are (spoiler: they are really, really damn stupid a lot complex though). Let's demystify them. A CPU is a bunch of logic gates which roll from one state to the next at every clock tick. We have not talked much about logic gates yet, let's talk about the NAND gate: When its inputs are all 1 its output goes to 0 otherwise it is 1. The magic is: If you have only that component, but lots of, some tens or hundreds thousands, you can make (working too hard to design all their connections) a CPU, thus a computer! All kinds of logic gates are actually used, they can make flip-flops that store one bit when the clock signal ticks (actually rises or falls), compare bytes, add bytes and <u>construct a big machine that fetches bytes from memory, find out which command they contain and do that command's action (execute it).</u> Then go to the next command and so on, doing that forever. What are those commands inside the memory forming the program? Is for example the command "play sound file "xxxx.xxx" among those? Nope. Is the command "open application xxx" one? Nope. Neither is the command "transmit xxx data over serial port". The set of commands more or less is:

- NOP: No OPeration, just a clock tick delay
- A set of "move" commands that copy a byte's or word's content of an address in memory to somewhere else. Other than memory, CPUs have a few internal bytes called CPU **registers**. Each has its name and is a "quick access" byte or word of internal memory (not RAM, internal to the CPU itself). A note here: in MCUs all peripherals' (GPIOs, UART, ADC, etc.) bits and bytes for configuring them and reading their data are called registers, each of those resides at a unique memory address. So in the case in ICs connecting over I^2C or SPI. Each of their bit or word functionality is explained in their datasheet.
- Add / subtract two bytes or words
- Multiply / divide (some CPUs do not have those! Addition and negative numbers can do all math calculations, really!)
- Do logical (called Boolean, will meet them next) operations (logic gates) AND, OR, XOR between two bytes' or words' bits
- Shift a byte's or word's content by x bit positions left or right
- Jump to x address (the next command to execute) (normally the next command to execute is the next command in the memory)
- Compare two bytes or words (numerically). Jump to x address if comparison result is ....
- Call x address: That is a jump returning back when the "return" command is met

And that is all about it! That stupid machine makes a Tesla car drive itself or implements a game simulating an ultimately realistic 3D environment. So... we need lots and lots of commands to do useful stuff and lots of those executing per each second. Here is the world of programming, beautiful and terrifyingly huge. It takes 100's millions of commands to make a smartphone with only one internet browser app in it. Fortunately we have made other "higher level" programming languages.

The language the CPU executes is the only real language of a computer, it is called **assembly language** or machine language or op-code or binary code or object code. Actually each machine language command is a specific number, e.g. NOP might be number 27, assembly language is an 1-1 representation of the word "NOP" to the number 27 in order to be "human readable".

Thankfully we have made programs that process other, better to the programmer commands, like "open x file" producing the machine language sequence of commands needed to implement their functionality. Those are "higher level" **programming languages** doing greater things per command (the machine is at the lowest level, its commands are dummy). At the old days there was only assembly. Now we have a toolbox of many great languages of very high level to program a lot easier.

They are divided in two kinds. Compilers and interpreters. A **compiler** takes a text file, checks it (textually) all for any syntax errors, if ok it converts it all to a machine language program that will function as intended by that x language. We then execute that "binary code". In the MCUs world it is transferred from our PC where the compiler is to our MCU by a serial (UART) port and a bootloader program in the MCU or directly using a hardware debugging protocol (JTAG, SWD etc) over a USB dongle. This downloading involves also writing it to the flash memory, we say we "flash" the MCU. An MCU resetting after that starts its execution. The term **source code** refers to our program in its x programming language as a text that is written by us. That is human understandable, and may also carry comments with it. The term **object code** refers to the binary code produced. It is terribly difficult for us humans to understand the object code in the assembly language in terms of functionality.

Interpreted languages are programs taking a text file containing our program (at the syntax of the x language) and execute it command by command. That is they analyze the first command, if ok they execute it (do the actions it should do), then they analyze the next command etc. They are way slower since they do this command analysis for every command at "run time" (usually around 20 times slower) but they are handier. Such languages are Python, Java, JavaScript, all very-very high level. They also need that language execution program to fit in an MCU that requires usually 5-20 times the memory of Arduino UNO's ATmega328. Historically the first home computers came with a medium level language interpreter called "Basic".

That all was about how the machine under the hood is. Let's move on to use them in understanding and learning the most important programming language, C and its extension, C++.

2.12 C++ INTRODUCTION FOR THE NON-PROGRAMMER

C++ is a version of C with awfully lots of extensions. C, this one letter name language was born in 1973 and since it is the de-facto language for making efficient (fast and small) programs. The biggest PC operating systems are still written in C. The charm of this language that makes it never wearing off is that it is low level (assembly language alike), so the resulting binary code from the compiler is small and fast if our programs are well structured. It also offers high level functionality and structured syntax. C++ is only "high level extensions" keeping all the low level as it is and compatible to the older C. The majority of the MCUs nowadays intentionally uses C, not C++, since C++ has the burden to be around 2 times slower. You cannot be both super functional and super-fast.

Arduinos have chosen C++ because the 99% of their mission is to make programming easier. C++ offers a lot of complexity and with it a lot of easiness in doing things. Arduino IDE hides illusively about 5 lines of code needed at every program that make it appear more complex and more tedious. They also present two function blocks of code to place our commands inside, setup() and loop() and a wealth of functions doing specific MCU functionality, almost the easiest possible way (e.g. digitalWrite(pin, state)). You may fall into the name "**Arduino language**". Well, it is not fair to call it a language, it is C++ with less than 1% of its stuff hidden from you but existing in the background. Arduino IDE also uses .ino instead of .cpp filenames for your (source) code, but they are just plain text files, just as the .cpp files are. Arduino's IDE programs are called **sketches** though they are just C++ programs.

We will start with C's basic stuff and then go to the extensions the C++ offers.

Numeric and data types:

There is not one kind of numbers. Since CPU can only handle integer numbers of constrained bit length (ATmega for example is 8 bits only), there are various kinds according to what we want to present numerically, from a single byte (0-255) up to big floating point types (e.g. 3.14159265359).

In general there are integers divided into signed and unsigned of 8, 16 and 32 bit length (e.g. 8 bit unsigned range from 0 to 255, signed range from -128 to 127). Remember that 16 bit holds $2^{16} = 65536$ values, and 32 bit hold around 4 billion. Besides integers there are the magic (but slower to calculate) floating point numbers divided into floats and doubles that is single precision and double precision respectively. Single precision hold from 3.4×10^{-38} to $3.4 \times 10^{+38}$ or written in the C language format, 3.4E-38 to 3.4E38 having about 6 "significant" digits e.g. 7.12345E6 might be one such number (the number 12.3uF could be written 12.3E-6).

Besides numbers there are character types also. How is text stored and manipulated? Each letter has a unique code called "ASCII" code, googling "ASCII table" you will find them and they are enough to fit in one byte (255 such codes). For example "A" is 65. Character types treat content as characters, we need printing text and manipulating text often in our programs. Here are all basic types:

- **boolean** (8 bit) - simple logical true/false
- **byte** (8 bit) - unsigned number from 0-255
- **char** (8 bit) - signed number from -128 to 127.
- **unsigned char** (8 bit) - same as 'byte'; if this is what you're after, you should use 'byte' instead, for reasons of clarity
- **word** (16 bit) - unsigned number from 0-65535
- **unsigned int** (16 bit)- the same as 'word'.
- **int** (16 bit) - signed number from -32768 to 32767. This is most commonly used for general purpose variables in Arduino example code provided with the IDE
- **unsigned long** (32 bit) - unsigned number from 0-4,294,967,295. The most common usage of this is to store the result of the millis() function, which returns the number of milliseconds the current code has been running
- **long** (32 bit) - signed number from -2,147,483,648 to 2,147,483,647
- **float** (32 bit) - signed number from -3.4028235E38 to 3.4028235E38. Takes 4 bytes and usually about 20 times longer to calculate than using integers.

Variables:

As in mathematics, variables are entities that contain some kind of a number, more specifically, a type of a number. Those entities get a name from us which most helpfully should represent their content. For example we may name a variable "speed" of type float containing the speed a car is moving in km/h. Since each variable should have a specific type, we have to declare its name and its type at some point (the beginning usually) of our program. Yes, we could be using only floats but that would take too long to execute in our 8-bit ATmega328. That is an example of the efficiency a "close to the machine" language like C provides e.g. when a variable may be just a counter from 0 to 10 we may use the byte type. In the following example we will show how a variable is declared (to its name and type) and how it is used

Code	Explanation
`int counter = 100;`	- a new variable
`float diameter;`	- a new variable
`float perimeter;`	- a new variable
`void loop()`	
`{`	
`   counter = counter - 1;`	Assignments
`   diameter = 20.2;`	
`   perimeter = diameter * 3.1415926;`	
`}`	

Language keywords (commands):

Amazingly they are very few. C has 32, about half of which are declarations of data types (like "int" for integer) that make no action C++ has 95 in total (including C's), most of which are for very advanced concepts. C is designed in a way that it will seamlessly "extend" its commands with limitless new made from us or others using the "function" concept. digitalWrite(pin, state) for example is a function written by the Arduino team which we use as if it was a language command. We can write a lot more of our own, our program is always structured in function blocks and function calls (execution of a function is done by "calling" the function). In C++ the code structuring is further enhanced by using "objects" made out of "classes". We will get to those later. We will also get later to the commands or keywords of C and a few of C++.

Language syntax:

To be honest, C and C++ syntax is kind of tedious at some issues, yet in general it is awesome.

All commands should be ending with a semicolon ";"! That sounds like a tedious thing but we find it even at very modern languages like Java and others. You get used to it very quickly. The compiler will alarm you in case you forget it. We are not using the ";" everywhere either, you will easily get used to that too.

A great structuring element is the code block that is code surrounded with "{" and "}". Being a structured language makes us understand complex programs (ours also) more easily. Let's see as a dummy example this small part of a program:

```
update_temperature();
if (temperature > 40.2)
{
    digitalWrite(10,HIGH);
    delay(10000);              // delay 10 seconds
    update_temperature();
    if (temperature < 39.5)
    {
        serial.println("OK");
    }
    digitalWrite(10,LOW);
}
digitalRead(11);
```

Each code block (inside an opening and closing brace) is purposely written one tab position to the right. Keeping this way of writing makes the code blocks and all the program structure more visible. You should understand what the previous part of a program is doing, guessing the syntax and function of the "if" command, disregarding any knowledge about the whole application. The execution of the code may enter the big "if's" block or may not, if it does, it may enter the inner "if's" code or it may not. Some syntax we should notice is: ";" is spared to commands (language keywords) followed by blocks of code such as "if". Comments of one line just have to begin with "//". Multiple line comments have to begin with "/*" and end with "*/". Keeping code of each block one position to the right is called "indenting" and must always be kept even if the language (the compiler) will not report any error if it is violated. Last note: All language code is case sensitive! "If" or "IF" will not work! Temperature and temperature are different variables!

Functions:

Functions encapsulate functionality that is repeatedly used or they just isolate some special functionality code. They receive input and provide a numeric output, hence they have data type. In case we do not want to output something or "return something" there is a special data type called "void" that is "nothing" to return. We will see the syntax and the usefulness through this example:

Code	Explanation
```void blink(unsigned long time, int pin){ digitalWrite(pin, HIGH); delay(time); digitalWrite(pin, LOW); delay(time);}```	- Implementation of a function called blink, returning nothing, taking as parameters two integer numbers
```int max(int num1, int num2){ if (num1 > num2) { return num1; } else { return num2; }}```	- Implementation of a function called max, returning an integer number, taking as parameters two integer numbers
```void loop(){ blink(100,5); blink(1000,9); blink(max(200,250),4);}```	- A function call- Another function call- And another two function calls

Here we made two functions, the blink() that returns nothing (is a void function) and the max that returns an integer. Return command (is a keyword) exits the function code (returns from the function's call) while assigning a value as a function's result. The number of parameters and their types can be any, even none at all. Such a function is impended like:

```
void blink(void)
{
    ...
    ...
}
```

We call this like: `blink();`

We have already got used to function calls from using the digitalWrite(...), delay(..) and the other functions we used so far from the Arduino library.

A very important aspect is that **all code that does actions is placed inside functions.** setup() and loop() where we place our first code are functions also. Out of functions may only be variable declarations or some few other declaration mechanisms of the C or C++ language, nothing that will do some action though. We can say that C is a heavily "function oriented" language (there is no such term officially). We are surely constrained to how we have to write our code, that is very much appreciated though in big programs where those constrains create a magnificent architecture.

If you feel dazed you are not stupid, C and C++ are moderate to hard in difficulty to learn but we surely can, and when we do we will do real hard-core programming. Learning just the 20% of it will make us 80% creative. Seat back and relax. Read again stuff you did not understand at first reading.

For recapping let's put all said in another now functional example. Using the "Basic I/O" project of chapter 2.10 (a button a potentiometer and 3 LEDs) let's make a software project that blinks all LEDs sequentially when the button is pressed shining at brightness adjusted by the potentiometer.

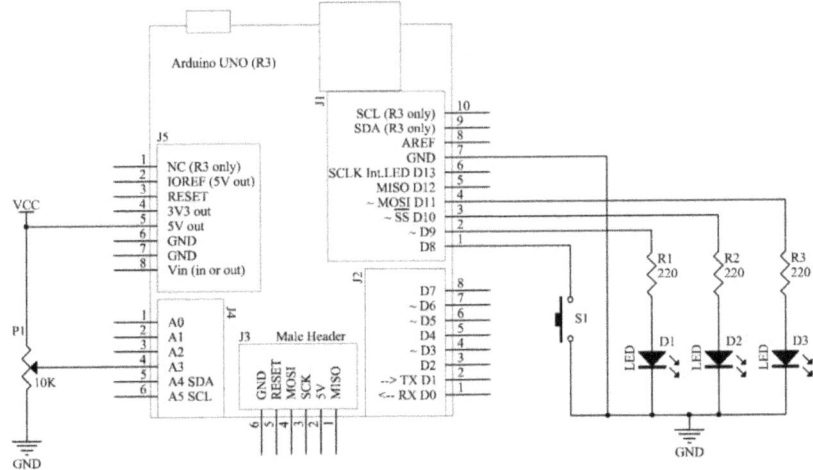

```
byte brightness;

void blink(int pin, byte duty_cycle)  // our blinking function
{
    analogWrite (pin, duty_cycle);  // PWM
    delay(500);                      // wait doing nothing for half a second
    analogWrite (pin, 0);   // 0 PWM is off
    delay(500);
}

void setup()   // this function executes once at startup
{
    pinMode(8, INPUT_PULLUP);
    pinMode(9, OUTPUT);
    pinMode(10, OUTPUT);
    pinMode(11, OUTPUT);
}

void loop()   // this function repeats executing forever
{
    brightness = analogRead(A3)/4;  //0...1023 -> 0...255
    if (digitalRead(8) == LOW)//while pressed it connects to ground
    {
        blink(9, brightness);   // calls blink(). Blinks D1 once
        blink(10, brightness);  // called when previous finishes
        blink(11, brightness);  // called when previous finishes
    }
}
```

Take your time to realize what this program does (e.g. pressing the button for 1.5sec, releasing it for 0.5 sec and pressing it again for 2 seconds) and how it is structured at all its details. (Answer: D1 blinks once, D2 blinks once, D3 blinks once and that sequence repeats one more time without in-between pause)

2.13 More Arduino programming

Having told about the structuring, data types and variables, we will dive deeper to the C and C++ language, always visiting the most practical places of this big tool-chest. Give your patience to the few pages of this chapter, the next chapter will be more playful, consisting of examples. We will cover the basics, this is not a full language reference.

Symbols for numeric operations (operators)

We will see the most useful ones arranged in an array since most are really simple.

Category	Symbol	Function	Example
Numeric calculation	=	Assign right to left	`var = 24;` `var1=var2*2;`

Numeric calculation	+ - * /	Add, subtract, multiply, divide. Note that * and / happen first, then + and - happen unless parenthesis are used	`var = 2+4*2;`
			`(result is 10)`
			
			`(2+4)*2`
			`(result is 12)`

Numeric comparison	==	Checks if equal	`if(var1==var2)`
Numeric comparison	!=	Checks if different (≠)	
Numeric comparison	<	Checks if left is less than right	`if(var < 10)`
Numeric comparison	<=	Checks if less or equal	
Numeric comparison	>	Checks if greater than	
Numeric comparison	>=	Checks if greater than or equal	
Abbreviations	+=	If we are to write var = var+5 which is common, we may write var+=5 instead, same for -= *= /=	`var+=5;`
Abbreviations	++	If we are to write var=var+1 which is common, we may write var++, same for --. Imagine what C++ means.	`var++;`

Regarding the comparison operators, we should mention than in C and C++ 0 is "false" or "fail", all other numbers are "true" or "success". You will see the TRUE and FALSE values in same programs, they are just equal to 1 and 0 respectively, just as is HIGH and LOW in pins states.

BOOLEAN LOGIC:

We left that out of digital electronics hardware. It is what logic gates do. It is very applicable in software only nowadays, so let's see the Boolean logic invented by Mr Bool at around 1850. It is about what one could do in a world of 1s and 0s. It is the most basic and essential ingredient in CPUs and MCUs hardware (in the chip). There are some operations we can do in the 0/1 world, the basics are:

Input 1	Input 2	AND	OR	XOR	Single input	NOT
0	0	0	0	0	0	1
0	1	0	1	1	1	0
1	0	0	1	1		
1	1	1	1	0		

In software they apply a lot in true or false logic states (remember, true is 1, false is 0) e.g. if one result AND another result is true. They also can apply to individual bits the same way. Operators are:

Category	Symbol	Function	Example
Boolean calculations for bytes or words (applied to all bits)	&	Calculates the logic AND of two numbers bit by bit	In binary: 0010 & 1110 equals 0010 or var = 2&14
	|	Calculates the logic OR of two numbers bit by bit	
	^	Calculates the logic XOR of two numbers bit by bit	
	~	Calculates the logic NOT of one number bit by bit	byte var; var = ~15; //(~00001111) (result:240)
Boolean calculations for two entities	&&	Calculates AND operation with inputs what is on its left and what is on its right	if(var1>20 && var1<100) calculates var1>20 (for true or false) then var1<100) and applies AND to results
	||	Calculates OR operation with inputs what is on its left and what is on its right	("|" is the symbol over "\\" in keyboards usually)
	!	Reverses (NOT operation) what is on tis right	if(!(var>10))

A note: Only when bits in a byte are to be manipulated, it is handy to use hexadecimal numeric system. Hex digits are 16 (6 more than decimal system's 10 digits) going from 0 to F (F=15) as 4 bits go also from 0 to 15, so memorizing those 16 bit patterns helps visualizing directly each bit's value of 4 bits groups. 0x4F for example (0x is the hexadecimal notation) is same as 1000 1111 in binary, also same as 4*16 + 15= 79 in our decimal system. Use hex as less as possible.

BASIC PROGRAM EXECUTION FLOW CONCEPTS

It is clear so far, but let's repeat it, that a CPU executes only one command of the program at a time. Multitasking on your PC happens by letting your CPU execute a program for a little while (msecs), then another one for a little while, then another, switching so fast that you are tricked feeling their actions happen concurrently. An operating system is required for doing this, not the case in Arduino UNO and most MCU applications. So, while our program is running (it executes) if we slow down the time we will see that it executes a particular command, then go to another, then to another, we can imagine there is an execution point wandering through or program's lines or commands. The path it takes follows different patterns like a circle or loop in the loop() function's contents. That is the flow of the execution and it is how the "machine" of our program works.

Function call: When a function is called, the execution point jumps to the start of the code in that function. When the return command is met it returns back to executing the next command (after the function call command)

if: Takes a condition as input, if it evaluates to true the execution point goes to the first command inside its code block, otherwise it jumps to the first command after it

```
if (var1 > 20 && digitalRead(8) == HIGH)
{
    command;      // execution goes here if condition is true
    command;
    ….
}
command;          // execution goes here if condition is false
```

Notice the syntax details, we will see this in many examples (as we already have)

if-else: Same as if with two blocks of code. Needs not further explanation other than its syntax:

```
if (condition)
{
    command;    // execution goes here if condition is true
    command;
    ...
}
else
{
    command;  // execution goes here if condition is false
    command;
}
```

So one of these blocks will execute and one will not.

while: It takes a condition in the same syntax as "if", followed by a block of code. What is does is, it **loops**. Looping in programs is the cyclic re-execution of a part of it. Exactly what while does is:

```
while (var1 > 20 || digitalRead(8) == HIGH)
{
    command;  // execution goes here if condition is true
    command;
    ...
} // When execution reaches here it goes back to the while()
command;  // execution goes here if condition is false
```

another example:

```
var1 = 10;
while (var1 < 15)
{
    command;
    var1++;
    command;
    ...
}
command;
```

will loop the while block 5 times leaving var1 equal to 15 at the end.

Programs execution flow must never reach an end, but commands execute at a rate of about 10 millions per second. The usual way to make that happen is to place the most of your code inside a while(1) while! In Arduino programming that is already done for us. loop() function is in such a **forever loop**, when it finishes and exits, it is called again, forever.

for: Like the last example, many times we want some program portion to loop for x times. "for" command is made for that. A variable has to be dedicated to it that will store the counter's value. "for" chose a rather complex syntax as to be more generic or have broader functionality, at its counting mode it is like:

```
int var1;
for (var1=0; var1<5; var1++)
{    // will execute this block 5 times
    command;
    command;
    ...
}
command;
```

So, inside the "for" parenthesis we meet an initialization then a comparison then an action per loop. for(;;) is exactly like while(1).

break; command exits the code block of the "while" or "for" upon its execution, for example:

```
var1 = 10;
while (1)
{
    command;
    var1++;
    if (var1 > 15)
        break;
    command;
    ....
}
command; //after execution of break execution point jumps here
```

Note that "if", "else", "while", "for" command's block of code can also be one command only. In that case we may spare (optionally) the braces since a block of code acts in the syntax as grouping its commands into one. Nice to use indentation even to that solo command since it belongs to the "if".

We are at about the 80% of the usefulness of the C / C++ commands that do actions (not declarations). We will stop here.

STRUCTURES AND ARRAYS

We will speak about those only in very sort, up to the 20% of their knowledge, just to get you into the spirit.

In many cases of bigger programs we have a lot of data thus a lot of variables. Structures are super-variables that contain other variables. In a program for example that is a game of moving cars we may make a "car" structure containing variables such as speed, km_covered, direction, lights_state,…. That structure will work as a variable type making it then easy to create a lot of car type variables each with its own data. We call each of those an instance of the structure. Having made instances we access each of their member **using a dot** like:

```
car2.speed = 24;
```

Arrays or tables are more common in use, especially the one-dimensional arrays. If for instance we measure an ADC input every second and we need to keep the last 100 measurements somewhere, making 100 variables should be terrible. We can make the array

```
int value[110];
```

that will provide 110 int variables (just a few more than needed to be sure our data will fit) accessed as:

`value[0]` for the first one, `value[1]` for the second, ... `value[109]` for the 110th. A variable of ours can be used to choose which one to use, or be the index of the array. Accessing `value[110]` or any higher index makes our cars and our program crash.

CLASSES AND OBJECT ORIENTED PROGRAMMING

C++ is actually the addition of object oriented programming in C. OOP is indented to hugely assist programming structuring, the bigger the program is. Well... in practice it finds it's meaning in programs bigger than 1000 lines. Arduino, having built its house on the foundations of C++ caries both its advantages and its complexity. In very sort here is the objects story:

A class is a "mold" producing objects. Classes are similar to variable structures with additions, one of which is that they also contain functions, thus code. That makes them like autonomous programs themselves. In our cars game, using classes enables us to add functionality to the car structure, so an object could also do:

```
car2.accelerate(10);
if (car2.speed > car3.speed)
…
```

Lets leave OOP here, it is about 3 times more lengthy knowledge than all C language's knowledge. You should just know that in dots there is a class behind and an already made for us object. A great portion of Arduino libraries is made that way.

SERIAL COMMUNICATION

This is crazily too useful. It is your door (the only one in Arduinos) to get yourself inside the data of your program and your MCU in general. It is about the Arduino's UART communication with a "UART over USB" on your computer, fortunately in Arduino UNO and others there is one "UART over USB" already on your Arduino board. Let's leave a picture to tell the most of the serial communication story. Read the program in there also.

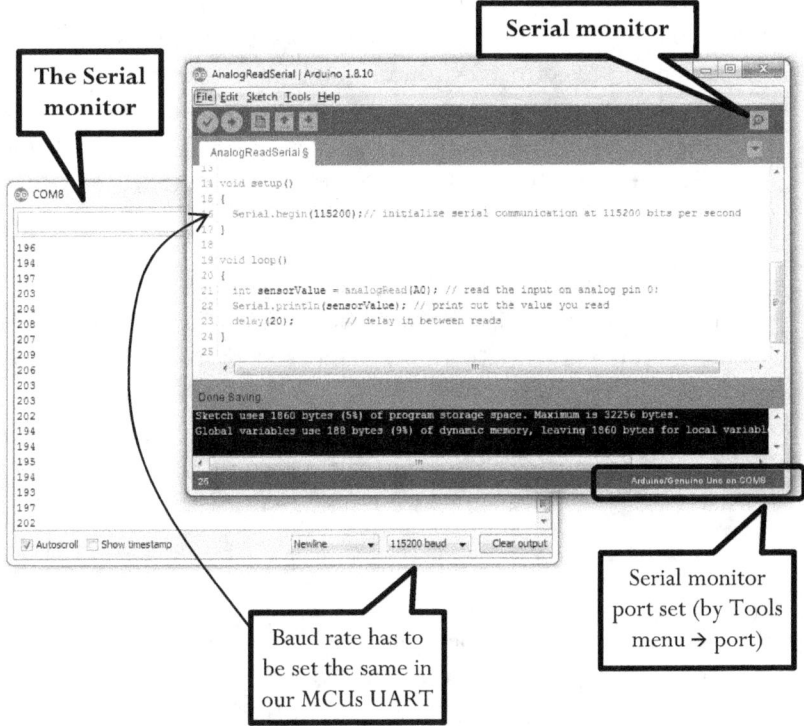

"Serial" is an object already created waiting for us to use it. Serial.println() function prints its parameter as text followed by an "Enter key pressing" to go to a new line. There are more function members of this class, also for reading from your UART if you send information over the serial monitor or from another application or any UART, like receiving data from a GPS device for example.

The serial monitor (or other serial terminal applications as well) have a great value on debugging your program.

DEBUGGING YOUR PROGRAM

Assume you made a mistake somewhere in your program and you notice it behaves oddly. The mistake is called bug, debugging is what you need to do. That is a case happening once every a few minutes when you are developing software. In order to find out what is wrong you need to see some data at some point of the execution. Same case is when you have made a small portion of your whole program and need to see if it does its job well. The way to do this with the Arduinos is to send information to the Serial Monitor over the UART in the fashion we did on the previous picture. (The story of the bug sort: In a 40's pioneering electro-mechanical kind of computer / calculator weighing 25 tons, trying to find why some calculation results were wrong they finally found a relay stuck by a small butterfly what made it its nest! A real story of hard times in the early computer history.)

STRINGS (TEXT MANIPULATION)

String in programming is not a piece of thread and it is not the sexy lingerie either. It is a "cord" of characters, actually a character array in C that contains text. Text therefor is char type arrays, ending with a zero (null-terminated). Lately there is an easier to learn and handle library called string (it is a class). It is better not to dive into details here, just giving you a picture and some titles.

SCOPE

Each variable may not be accessible at all parts of our program. The rules are simple. If it is created within a block of code (functions' body, inside a for, a while etc) it is not "visible" outside of it. When the execution exits it's block of code it automatically "dies", it disappears and the information of its value is forever lost, unless it is declared with the extra keyword "static". You may meet variables defined (created) inside code blocks (like in our last program for the serial example). It is preferred to do that outside of all blocks in simple programs. Creating variables inside loop() especially involves a trap. loop() actually exits when its last "}" is met and then it is called again. That resets any variables created within it.

THE INFORMATION NOT COVERED YET

An awesome place to find any further information to almost all its specifics is in the great arduino.cc web site, at the Arduino reference page. Just "google" arduino reference. 10000% recommended.

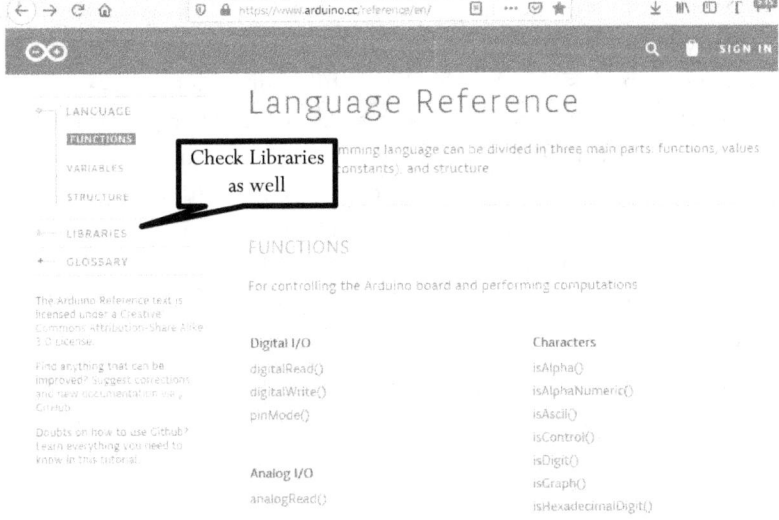

2.14 A FEW SIMPLE PROGRAMS TO PLAY WITH

It's time to apply all this theory to practice. We will also seize the opportunity to add-up just a few more Arduino software related knowledge. To get used to programming or to a new language you need examples and mostly the joyful feeling "let's play". Here we go, step by step. Take time to read all those programs at their finest details, picture in you head all they do and imagine tampering them. To learn programming you need to have fun with it and make it the purpose, not the means to achieve the purpose. Experience tells you can never be great at something if you don't like or love it. Such feelings may develop only by "doing", now you are in the "reading" phase.

For hardware we will be using the project#2 of 2.10:

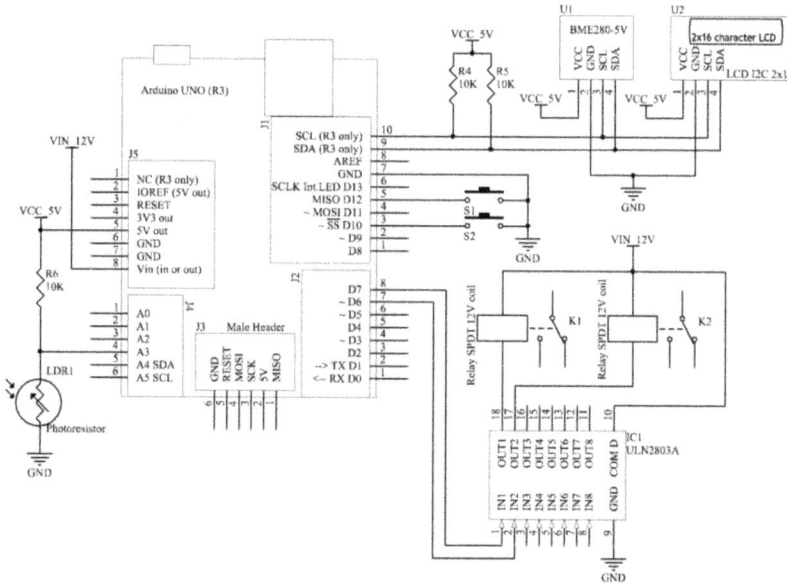

A MORE ADVANCED "BLINKY"

In all programming languages it is a tradition to introduce for the first and simplest program one that prints on the screen the text "Hello world!". In MCUs that first and simplest program traditionally is the blinking of an LED or just the periodic toggling of a GPIO. In our case we already have seen that. Let's see it again using an LED we do not have to connect, it is already placed on the Arduino UNO board, connected internally to the GPIO 13 as you can see on its schematic (in 2.9). So this program functions with nothing connected to our UNO.

Here it is:

```
void setup()
{
   pinMode(LED_BUILTIN, OUTPUT);
}
void loop()
{
   digitalWrite(LED_BUILTIN, HIGH);   // turn the LED on
   delay(1000);                        // wait for a second
   digitalWrite(LED_BUILTIN, LOW);    // turn the LED off
   delay(1000);                        // wait for a second
}
```

LED_BUILTIN equals to 13. This is set inside the Arduino libraries using the command: (we have to open a specific file to see that)

`#define LED_BUILTIN 13`

If e.g. we use at the beginning of our program (out of all functions):

`#define ANSWER_FOR_EVERYTHING 42`

At any place we use ANSWER_FOR_EVERYTHING the language will replace it with 42. The use is to change all those 42 values used at many places changing only one at the beginning. "#" starting commands are commands of a text-preprocessor needing no ";" at the end. Later on we will see another one, the `#include`

That simple and good looking program has this problem: for each whole second of time the function `delay()` is executing, the MCU does nothing else, and usually that is bad since actually our program "freezes" for that time. `loop()` loops once every 2 seconds.

The following program loops its loop() function many thousand times per second, doing the same action on the LED.

```
unsigned long time_LED_toggle;   //initial value is zero
byte LED_state;                  //initial value is zero

void setup()
{
  pinMode(LED_BUILTIN, OUTPUT);
}

void loop()
{
  if (millis() - time_LED_toggle > 1000)
  {
    if (LED_state == 0) // LED was OFF
    {
      digitalWrite(LED_BUILTIN, HIGH);
      LED_state = 1;
    }
    else  // LED was ON
    {
      digitalWrite(LED_BUILTIN, LOW);
      LED_state = 0;
    }
    time_LED_toggle = millis();
  }
  // we can add more code after here doing other things
  // ...
  // ...
}
```

millis() returns the time in milliseconds since program start. **The technique is to measure time and according to "how much time has passed" to decide if actions will be taken.** Each C++ command or call of a simple function like the ones used here, usually takes less than 100 clock ticks of our MCU's clock to execute. In the kind of slow MCU of Arduino UNO, clocked at 16MHz, that time is around 6usecs or around 150,000 such instructions per second. Getting inside the "if" or not, the execution point passes really fast through all the LED handling procedure. Take some time to understand it to the last detail. A *note for those more advanced in programming: If someone wonders about what happens when the 32 bits unsigned long type of* millis() *function overflows (wraps to zero when filled over the value 2^{32}=4294967296 after 4294967.296 seconds that is 49.7 days),* *subtraction of numbers treated as unsigned 32 bit integers in that program will not even produce the slightest glitch. A version like:* `if (millis()>time_LED_toggled+1000)` *would be buggy in that matter.*

In programming it is very seldom to have only one way of achieving a goal. The previous program was intended to present a very commonly used method. You may imagine others also. One might be to turn the LED on if the modulus (the remainder of the division) of the `millis()` to the number 2000 is less than 1000, off otherwise (In C, that calculation operator is the "%" or `millis() % 2000`). Last, there are many ways to implement a specific method, all you have to do is test the code you write in your Arduino to see how well it works. The method taken to implement a functionality or solve a problem (e.g. we will act according to what time it is now and do that or that) is also called **algorithm**. Program structuring for implementing a method (e.g. encapsulate our code inside a new function) is not an algorithm, it is just coding practices or code structuring.

UP AND DOWN WITH TWO BUTTONS

The mission: The user will adjust a setting temperature using the up (pin 12) and down (pin 10) buttons. Before displaying that adjustable setting on the screen we should calculate it properly (according to buttons actions) as the value of a variable. Using a variable for this purpose is 100% correct and the only way to do it. So, before starting to play with the screen lets prepare our internal data.

Variables should have a proper name to make us and others understand what they contain. Don't use a name "t" for this, let's name it `temperature_set` choosing float type since it is common to have fractions of a degree. Let's work on Celsius units only.

Until we setup our LCD screen we will use the classic method to see a variable's value, live, while our program is executing. The serial monitor. `Serial.println(temperature_set);` will execute many times per second in the loop()'s loop.

Using an algorithm like "if button is up wait forever (checking its state) until it goes down, when that happens do something" should block all our program and the functioning of the other button. Here is one of the many ways to do this job along with the LED blinking:

```
unsigned long time_LED_toggle;    //initial value is zero
byte LED_state;                    //initial value is zero
float temperature_set = 25.0;     //initial value
byte button_up_previous;          // the previous state (0 or 1)
byte button_up_pin = 12;          // for easier moving to other pin
byte button_down_previous;        // the previous state (0 or 1)
byte button_down_pin = 10;        // for easier moving to other pin

void setup()
{
  pinMode(LED_BUILTIN, OUTPUT);
  pinMode(button_up_pin, INPUT_PULLUP);
  pinMode(button_down_pin, INPUT_PULLUP);
  Serial.begin(115200);
  // Initialize the buttons' previous states
  button_up_previous = digitalRead(button_up_pin);
  button_down_previous = digitalRead(button_down_pin);
}

void loop()
{
  // LED blinking
  if (millis() - time_LED_toggle > 1000)
  {
    if (LED_state == 1) // LED was ON
    {
      digitalWrite(LED_BUILTIN, HIGH);
      LED_state = 1;
    }
    else  // LED was ON
    {
      digitalWrite(LED_BUILTIN, LOW);
      LED_state = 0;
    }
    time_LED_toggle = millis();
  }

  // temperature_set by buttons
  if (button_up_previous==HIGH && digitalRead(button_up_pin)==LOW)
  {
    /* up button just pressed
    (remember volatge is 5V when unpressed, 0V when pressed)*/
    temperature_set += 0.5; // the increment at each button press
  }
  if (button_down_previous==HIGH && digitalRead(button_down_pin)==LOW)
  {
    // down button just pressed
    temperature_set -= 0.5; // the decrement at each button press
  }
  button_up_previous = digitalRead(button_up_pin);
  button_down_previous = digitalRead(button_down_pin);

  // we can add more code after here doing other things
  Serial.println(temperature_set);
}
```

Take all the time to read this over and over and understand its method. Comments help greatly for this by explanations inside the code. We have actually introduced 5 new variables, 3 necessary and 2 optional for greater flexibility. The concept is that we store at the proper place of the execution point's looping the previous or historic state and that enables us to detect when a change has occurred. We detect the "pushing event" of the button that is the falling edge of the GPIO's input reading (internal pull-up resistor makes it 1 while not pressed and 0 while pressed) and act only then.

INTERRUPTS

Another way to do this kind of functionality is using interrupts. In a nutshell, interrupts are a hardware mechanism of MCUs that automatically call pre-defined functions (interrupt service routines, ISRs) on the happening of an event. Since the happening of the event is checked by hardware, our program will jump to service that no matter what it does (where its execution point is) at that moment (delaying around 1-3 CPU clock ticks only!). Using interrupts is more complex. It tends to unleash a swarm of bugs to eat our program if we are not well educated about them. Interrupt receiving GPIO pins are only pins 2 and 3 on UNO, capable to "listening" to rising or falling edge or both, if configured. Keep them in your mind for monitoring events changing faster that key presses, where we must not miss any. *Even changing to those pins and using interrupts in our case, other problems occur, such as "key bouncing" that is, you may press the button once, but electrical noise when the two metals of the switch just start touching each other and the extremely fast responding of interrupts makes our program think we pressed it 1 to 10 times in a row (a random count of repetitions).*

LIBRARIES

Moving on we must grab more tools from a tool chest called "Arduino libraries". This is about integrating pieces of code into one program, pieces either ours or written by others.

C and C++ programs span on multiple files. That is for a great list of reasons, such as, scrolling up and down a program of 1 million lines should be a headache, code is better organized by this encapsulation by separating into files, files are included in our program easily. Arduino framework is a bunch of C/C++ files (more than 20!) silently included in our code. In those is the implementation of the `digitalRead()` function for example, the `delay()`, `millis()` and all other functions we have been using so far. Adding functions, classes and other C++ goodies, written by great people who give them for free or written from ourselves we expand the repertoire of functionality. The flow of the execution point of our program is driven only by us, we may call some of those functions, we may not, as it fits.

We will soon need the functionality of a character LCD screen controlled over the I²C bus. In order to set it up and only to print on it the letter "A" the code is more than 50-100 lines handling all internals of the LCD controller and an I²C interface chip. The datasheets of both are around 50 pages to read. Also we should read the datasheet section of the MCU's I²C peripheral and do all the setup it requires by tempering registers bits. Here is the "Arduino" way:

Step 1: Find our great library and fetch it from the internet.

In our case for driving our screen (literally it is a "driver" but not in an operating system), we do the following:

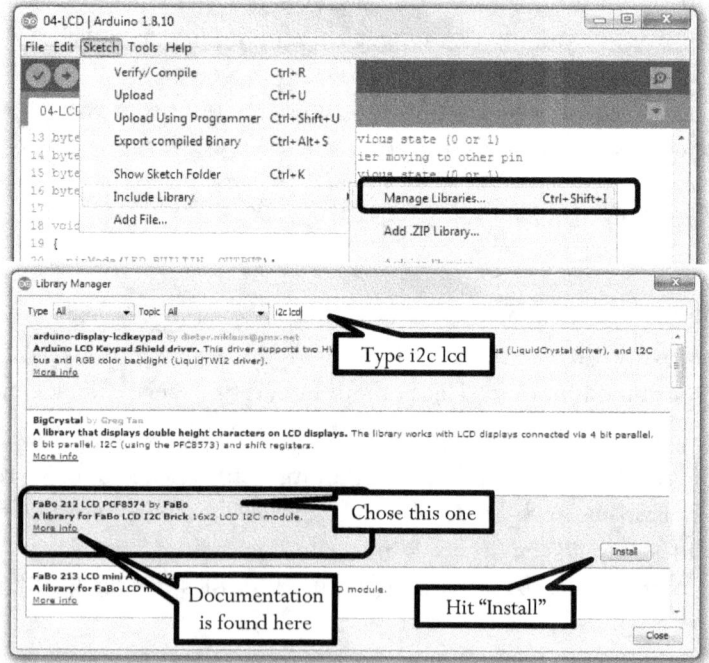

Step 2: Include it in our program: We go to sketch menu and choose it from the list of libraries. That will only add at the beginning of our program the pre-processor command:

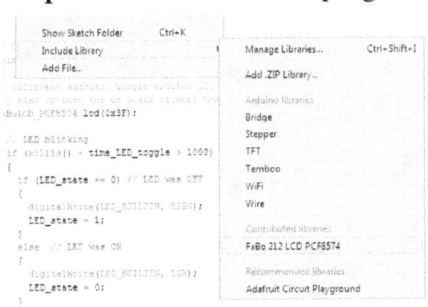

#include <FaBoLCD_PCF8574.h>

Where FaBoLCD_PCF8574.h is a filename. C++ files are of two types, headers ending in .h and sources ending in .cpp (.c for C). Header files are designed to be included with the #include command. Having done that we have all that library's classes and functions available to use. A final notice is that if we want to give our program to somebody else, we have to do the same procedure on her/his computer or copy the library folder from our "sketchbook location" folder to hers / his. That folder's location is set in the "preferences" of the "file" IDE's menu. In general Arduino IDE tries to involve you with C++ matters as less as possible in order for a beginner to feel the creativity joy ASAP. That is good but after a few projects you have to dig deeper to go on. Let's use this library...

DISPLAYING ON AN I²C SCREEN

Here is the code that displays the live value of temperature_set on the second line of the screen:

```
#include <FaBoLCD_PCF8574.h>

/* Create lcd object out of the FaBoLCD_PCF8574 class
using an initializing parameter that is the I2C address
If your PCF8574 I2C LCD does not work, it might be set
to different address. Google arduino i2c address scanner
Try also to turn the onboard trimmer that sets the contrast*/
FaBoLCD_PCF8574 lcd(0x3F);

void setup()
{
```

```
    button_up_previous = digitalRead(button_up_pin);
    button_down_previous = digitalRead(button_down_pin);
    // LCD setup
    lcd.begin(16, 2);
}

void loop()
{
    // LED blinking
    if (millis() - time_LED_toggle > 1000)
    {
        if (LED_state == 0) // LED was OFF
        {
            digitalWrite(LED_BUILTIN, HIGH);
            LED_state = 1;
        }
        else // LED was ON
        {
            digitalWrite(LED_BUILTIN, LOW);
            LED_state = 0;
        }
        time_LED_toggle = millis();
    }

    // temperature_set by buttons
    if (button_up_previous==HIGH && digitalRead(button_up_pin)==LOW)
    {
        // up button just pressed
        (remember voltage is 5V when unpressed, 0V when pressed)*/
        temperature_set += 0.5; // the increment at each button press
    }
    if (button_down_previous==HIGH && digitalRead(button_down_pin)==LOW)
    {
        // down button just pressed
        temperature_set -= 0.5; // the decrement at each button press
    }
    button_up_previous = digitalRead(button_up_pin);
    button_down_previous = digitalRead(button_down_pin);

    Serial.println(temperature_set);

    lcd.setCursor(0,1);
    lcd.print("Set point: ");
    lcd.print(temperature_set, 1);
    lcd.print("C");
}
```

Each library has its own set of functions generally called Application Programming Interface or API, like the Arduino has in all the inherent functions we are already using. Our programs have a hidden `#include <Arduino.h>` in them. Documentation of each API is found on the library's code repository (in "github") and on web sites if any. Last resort is to read a little bit the .h and the .cpp files of their source code. Note that making the above program without the benefit of an open-source well working library, it should take to write an equivalent of the `FaBoLCD_PCF8574.cpp` yourself that is extra 300 lines of quite complex code.

READING AN I²C SENSOR

Likewise, we are going to include the library SparkFun_BME280_Arduino_Library searching for sparkfun BME280 in our library manager. Here is one of the many ways to read the temperature and humidity:

```
#include <SparkFunBME280.h>

BME280 myBME280;   // Create an object out of BME280 class
float temperature;
float humidity;

void setup()
{

    // BME280 setup
    myBME280.setI2CAddress(0x76); // Try address 0x77 if next check fails
    // next the function myBME280.beginI2C() is called and its result is checked
    if (myBME280.beginI2C() == false) //Begin communication over I2C
    {
        Serial.println("The sensor did not respond. Please check wiring.");
        while(1); //Freeze
    }
}

void loop()
```

```
    lcd.print(temperature_set, 1);
    lcd.print("C");
  }

  temperature = myBME280.readTempC();  //see examples in library's
documentation
  humidity = myBME280.readFloatHumidity();  // we may also read pressure
and altitude
  lcd.setCursor(0,0);
  lcd.print("T:");
  lcd.print(temperature,2);
  lcd.print("C RH:");
  lcd.print(humidity,0);
  lcd.print("%");
}
```

```
T:18.99C RH:55%
Set point: 25.0C
```

Our code is a long train... let's organize it better.

A LITTLE CODE CLEANUP

The better the code structuring the less bugs we make, the quicker and happier we program. "Code refactoring" is also coined to converting a messy code to a cleaner one. We will only use functions here. Note that many coders take this organizing very far, making code 2 or 10 times as big for the sake of structuring, generalization and expandability. When you know where your project's functionality will end it is good not to stretch the expandability of your code (in terms of structuring) way far over that. A messy spaghetti code is worse of course.

We will just lighten the loop()'s lengthy code.

```
#include <SparkFunBME280.h>
#include <FaBoLCD_PCF8574.h>

BME280 myBME280;  // Create an object out of BME280 class
float temperature;
float humidity;
/* Create lcd object out of the FaBoLCD_PCF8574 class
using an initializing parameter that is the I2C address
If your PCF8574 I2C LCD does not work, it might be set
to different address. Google arduino i2c address scanner
Try also to turn the onboard trimmer that sets the contrast*/
FaBoLCD_PCF8574 lcd(0x3F);
unsigned long time_LED_toggle;  //initial value is zero
byte LED_state;                  //initial value is zero
float temperature_set = 25.0;    //initial value
byte button_up_previous;         // the previous state (0 or 1)
byte button_up_pin = 12;         // for easier moving to other pin
byte button_down_previous;       // the previous state (0 or 1)
byte button_down_pin = 10;       // for easier moving to other pin

void setup()
{
  pinMode(LED_BUILTIN, OUTPUT);
  pinMode(button_up_pin, INPUT_PULLUP);
  pinMode(button_down_pin, INPUT_PULLUP);
  Serial.begin(115200);
  // Initialize the buttons' previous states
  button_up_previous = digitalRead(button_up_pin);
```

```cpp
  button_down_previous = digitalRead(button_down_pin);
  // LCD setup
  lcd.begin(16, 2);
  // BME280 setup
  myBME280.setI2CAddress(0x76); // Try address 0x77 if next check fails
  // next the function myBME280.beginI2C() is called and its result is checked
  if (myBME280.beginI2C() == false) //Begin communication over I2C
  {
    Serial.println("The sensor did not respond. Please check wiring.");
    while(1); //Freeze
  }
}

void loop()
{
  LED_blink();   // LED control

  buttons_handler();   // temperature_set up-down using buttons

  // measure
  temperature = myBME280.readTempC(); // see examples in library's documentation
  humidity = myBME280.readFloatHumidity();  //we may also read pressure and altitude

  LCD_update(); // update content of LCD
}

///////////////////////////////////////////////////////////////

// LED control
void LED_blink(void)
{
  // LED blinking
  if (millis() - time_LED_toggle > 1000)
  {
    if (LED_state == 0) // LED was OFF
    {
      digitalWrite(LED_BUILTIN, HIGH);
      LED_state = 1;
    }
    else  // LED was ON
    {
      digitalWrite(LED_BUILTIN, LOW);
      LED_state = 0;
    }
    time_LED_toggle = millis();
  }
}

// temperature_set up-down using buttons
void buttons_handler(void)
{
  // temperature_set by buttons
  if (button_up_previous==HIGH&&digitalRead(button_up_pin)==LOW)
  {
    /* up button just pressed
    (remember voltage is 5V when unpressed, 0V when pressed)*/
    temperature_set += 0.5; // the increment at each button press
  }
  if (button_down_previous==HIGH&&digitalRead(button_down_pin)==LOW)
  {
    // down button just pressed
    temperature_set -= 0.5; // the decrement at each button press
  }
  button_up_previous = digitalRead(button_up_pin);
  button_down_previous = digitalRead(button_down_pin);
  Serial.println(temperature_set);

}
```

```
// update content of LCD
void LCD_update(void)
{
  lcd.setCursor(0,1);
  lcd.print("Set point: ");
  lcd.print(temperature_set, 1);
  lcd.print("C");

  lcd.setCursor(0,0);
  lcd.print("T:");
  lcd.print(temperature,2);
  lcd.print("C RH:");
  lcd.print(humidity,0);
  lcd.print("%");
}
```

Read it through once again. See how much better the loop() looks like.

A note for people knowing C++ language: To be correct in C/C++ language, our program should begin like:

```
#include "Arduino.h"
void LED_blink(void);
void buttons_handler(void);
void LCD_update(void);
```

Arduino IDE does this for us automatically. This is almost all the difference of the so called "Arduino language" to C/C++ language

STORING DATA PERMANENTLY

You should notice that Arduino stores no data when you turn it off. That is the problem of the RAM memory that is whipped out when left with no power. Flash memories hold their content practically forever. So your program can use a portion of the MCU's flash (32KB on ATMega328) to store stuff like the `temperature_set`'s value. That is 4 bytes (the data size of a float).

Here are the engineering constrains on that:

- Flash is divided into "pages" (128 bytes each on ATMega328). Changing even one bit requires to erase a whole page and re-write it.
- Writing a page in the flash takes some time (msecs) while MCU freezes its operation.
- Unfortunately Flash memory cannot be written many times. It wears off and becomes un-reliable after about 10000 recordings (referred individually to each page).

Fortunately we will not flash our AVR so many times with new program (fixing code retries) into it. That is a guaranteed limit, writing 10001 times should have a chance like 0.0001% to start malfunctioning. Datasheet on first page highlights: "Write/Erase Cycles: 10,000 Flash/100,000 EEPROM"

What is EEPROM? ATMega328 has 1Kbyte of it. It stands for Electrically Erasable Programmable Read Only Memory and can be written byte by byte. There is a library for it, the EEPROM library that you can find in the "sketch" → "Include library" menu without fetching it from the internet. That, like all the others which are already there are the **"internal Arduino libraries"** which you can read about in the reference page of Arduino. Arduino.cc web site is awesome there! Usage can be as easy as:

`EEPROM.put(0,temperature_set);` for recording (library takes care to write only if data is different, so not wasting the 100,000 recording cycles and `EEPROM.get(0,temperature_set);` for reading it.

THE CONTROL, THE REAL APPLICATION

The coding of MCU projects usually has this paradox: The real algorithm of it may be as little as 2 to 20 lines while all the project may be over 200 lines for providing user interaction, setup, communication, etc. Counting all the libraries code underneath, that ratio gets actually terrible. Let's get into the heart of this project, its real "control" part or "application" part.

Purposely we will keep a little vague or configurable by you what this circuit actually does since we learn fishing, not serving one fish dish only. For this let's only deal now with it being a thermostat, that is do what most heating devices in your house do: heat up until temperature reaches a set point, target reached, stop heating, if temperature drops under a point, re-start heating and forever loop. Let's also blink the LED while the temperature has not reached its set point yet (is not near it) and keep it always on while it is there. We will be turning relay K1 on when we need to heat it up, off when we do not need to stop heating.

A quite easy way to do that is the following. We will present only `loop()` here, in the rest of the program we have only added `pinMode(7, OUTPUT);` in `setup()`.

```
   pinMode(7, OUTPUT);
}

void loop()
{
   buttons_handler();    // temperature_set up-down using buttons

   // measure
   temperature = myBME280.readTempC();
   humidity = myBME280.readFloatHumidity();

   LCD_update();  // update content of LCD

   if (temperature < temperature_set)
   {
      // We have to heat up. Turn ON relay K1 on pin 7
      digitalWrite(7, HIGH);
      LED_blink();   // LED control
   }
   else
   {
      // Temperuture high enough. Turn OFF relay K1 on pin 7
      digitalWrite(7, LOW);
      // keep LED on
      digitalWrite(LED_BUILTIN, HIGH);
      LED_state = 1;
   }
}
```

Finally we will make our thermostat a little better, working like the "stupid" mechanical thermostats work:

```
void loop()
{
  buttons_handler();   // temperature_set up-down using buttons

  // measure
  temperature = myBME280.readTempC();
  humidity = myBME280.readFloatHumidity();

  LCD_update(); // update content of LCD

  if (temperature > temperature_set + 0.5)
  {
    // Temperuture high enough. Turn OFF relay K1 on pin 7
    digitalWrite(7, LOW);
  }
  if (temperature < temperature_set - 0.5)
  {
    // We have to heat up. Turn ON relay K1 on pin 7
    digitalWrite(7, HIGH);
  }

  // blink the LED accordingly
  if (temperature < temperature_set)
  {
    LED_blink();
  }
  else
  {
    digitalWrite(LED_BUILTIN, HIGH);
    LED_state = 1;
  }
}
```

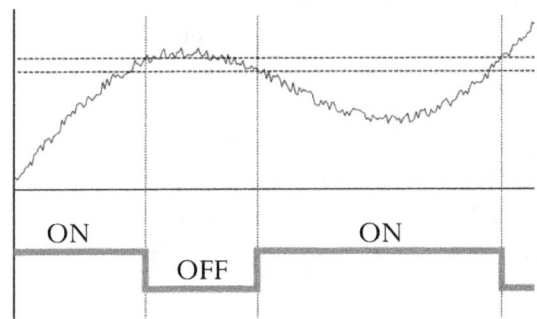

On the left we see what the two first 2 ifs do. The feature is called hysteresis. It solves the problem of switching the K1 relay on and off rapidly while the measured temperature is equal to the set temperature but the temperature's measurement noise makes K1 to "jitter". Do not say to yourself "oh, that is difficult to think and make". I made myself 5 tries to make it work. You have the luxury to experiment indefinitely, there is no software development "on paper".

2.15 WIRELESS COMMUNICATIONS AND INTERNET ENGINEERING

WIRELESS COMMUNICATION

We all wish we had a world without wires, but that is never going to happen since wires also carry the supply voltage. But it is nice to have a few remotely operating devices connecting to our phone or communicating with each other wirelessly. Here is where the technology, in low cost and practical applications, is at the moment:

Frequencies: Unfortunately the 99% of the frequencies are reserved for commercial (e.g. cellular telephony, TV, radio, etc) government, avionics, military and other uses. There are 3 major free bands: 433MHz (1.74 MHz bandwidth), 868 MHz (Europe), 915 MHz (America) (26MHz bandwidth) and the most common since it is used by WiFi and Bluetooth 2.4GHz-2.5GHz. Transmission on those frequencies is also limited in power and in duration (reasonable use makes them available to all).

Range: The range of communication is usually around 50 meters and in some rare cases of amplified devices it can go up to around 1Km in open field. Lately there are some ultra-long multi-Km range protocols like "Lora" but offer very low bit-rate in data throughput. The higher the frequency the more is the attenuation of the signal by obstacles such as buildings walls, even human bodies. "Line of site" range is usually 3 -10 times greater than indoors range. You will meet the term (xx)db (decibels) of transmitting power, (xx)db of receiver sensitivity. Their total (counting RX dbs as positive) makes the "link budget". Around 100dbs link budget at 2.4GHz makes 100m outdoor line of site range, doubling for every 6 dbs higher that it gets (e.g. 118dbs is about 100x2x2x2m=800m range), those said in great approximation.

Protocols: Two wireless devices need to be on the same frequency and "speak" the same communication protocol in order to "talk" with each other. Mostly known protocols are WiFi and Bluetooth (classic and low energy called BLE), other less known are Zigbee, z-wave, Ant, Lora and many other proprietary of various manufacturers used only by specific chips. Speaking of protocols, we assume data is exchanged, so MCUs are in the loop. Almost all such protocols are bi-directional, one can both send and receive. That way they usually acknowledge if each transmitted packet of information has been received well by the other side. Each protocol offers different functionality, data throughput and power consumption.

Wireless transceiver ICs: The word transceiver comes from the abbreviation of transmitter/receiver. Most of those communicate over SPI protocol with any MCU and handle all the functionality to transmit and receive radio waves. That is also called the "RF section", RF for Radio Frequencies. Others, especially using Bluetooth or WiFi include an MCU inside them.

Antennas: The higher the frequency the smaller the Antenna. Either small ceramic material antennas are used or no-cost antennas formed by is a trace of PCB with a geometry that tunes to the frequency, called PCB antennas. They occupy about 1cm x 2cm PCB area in 2.4HGz. External antennas 10-20cm long are also used but in practice very rarely. Any material placed near an antenna drops the signal a lot (at less than 1cm especially even non-conductive materials like PCBs or plastics make a nasty effect).

Notable RF transceiver ICs:

- NRF24L01+: One of the oldest (but still very good) on the market transceiver chips by a Norway company called Nordic. Needs 3.3V to 5V converter for Arduino. Make your research about fake NRF chips on the market. NRF24L01+ with "PA+LNA" offers real 1Km range. Cost of the standard (about 50m range) is around 0.7$!
- ESP8266 based boards like the "NodeMCU": Choose ones with lithium battery input socket. We have already described the unbelievable MCU's capabilities for about 2$ cost. WiFi is functioning awesomely offering an unbelievably long range.
- ESP32 based boards with most notable the ESP32 development kit or others with lithium battery input socket: We have already described the unbelievable MCU's capabilities for about 4$ cost.

- UART over Bluetooth modules: Offer really easy serial communication over Bluetooth. Note that Bluetooth classic (v2) and Bluetooth low energy or BLE (v4+) are different protocols.

LOW POWER DEVICES

What makes a wireless device wired is its need to connect with a supply. Using a battery to make it portable, the quest is to have as small battery as possible to make it operate for practically long periods of time. That is all about low power designs equal to low consumption current designs. There are low power MCUs out in the market offering "sleeping" modes consuming less than 3uA with a timer like an RTC running that is setup to periodically "wake up" the MCU. In that sleep mode there is another "current hungry" device in our circuit we have to deal about, the regulator. All battery operated circuits practically require at least one regulator. When an LDO's output is connected to nothing, i.e. no current is output from it but its output voltage is generated, it consumes current to operate itself. That spec is the "quiescent" or "ground pin" current. Most regulators have 1mA – 20mA quiescent current, but some specialty have 5uA or less!!

So, let's use a low power MCU like STM32L031F6P6 from ST costing around 1$ (LCSC.com) offering peripherals and memory just above those of the ATMega328 but much higher speed and better ADC. Its sleep current is 0.6uA. Let's use the LDO MCP1700T-3002E/TT from Microchip offering 1.6uA quiescent current 3V output with ultra-low dropout costing about 0.3$ and an NRF24L01+ (or NRF24L01P) for RF offering a sleep mode of 0.9uA. Here is an example of how we calculate our battery size and operating duration:

Assume we use a 2AH lithium 3.6V non rechargeable AA size battery. In sleep mode total consumption current will be 0.6uA + 1.6uA + 0.9uA = 3.1uA but let's consider that to be 4uA for taking some engineering margins into account and any unforeseen small but real current leakage due to parasitic resistances. Never waking up our battery will consume 4uA having 2AH energy reservoir, lasting for 2AH/4uA = 500,000 hours. One year is 8760 hours and as a NASA engineer said in a speech, there is the "engineering year" that is 10,000 hours, easy for calculations with margins accounted. Our battery will last for 50 years then! (not accounting for its self-discharging over the years).

If our MCU turns on for 3 milliseconds every 3 seconds to measure and calculate something sleeping at the rest of the time, it will consume about 5mA running at 32MHz for that period only, making an **average** current of 5mA/time ratio = 5mA/1000 = 5uA totaling our average current to 4uA + 5uA = 9uA. If our transceiver chip operates averagely every 1 minute (e.g. to transmit temperature measurement only when it has changed significantly) consuming 40mA for 5msecs it will consume **averagely** 40mA/time ratio = 40mA / (60/5msec) = 40mA/12000 = 12uA. Totally our device will averagely consume 4uA + 5uA + 12uA = 21uA lasting our battery for 2AH/21uA = 95238 hours = around 10 years! Using a classic size coin cell CR2032 of 200mAH would last for 1 year. The smaller the energy the smaller the device.

Note: the free "Console Calculator" by ZoeSoft enables you to type "200m/21u" and get the result. It's a tiny "must have" application on a PC or MAC.

That example was given to understand that low energy takes sort operating times and low quiescent and sleep current. It is also given to show how simple the calculations are. Bluetooth version 4+ that is the BLE for "Bluetooth Low Energy" is an example of this technique with many existing single chip 32bit MCU+RF ICs to do that, also compatible to smartphones! Protocols like WiFi designed for constant data streaming or others with very low bit-rate needing a lot of time to make a single transaction and sleep again, are not fit for low energy applications.

The Trend of "Smart" Devices

7x7mm sized MCU + radio BLE ICs have enabled the making of devices with a powerful MCU and a few sensors in the size of a keys tag (mostly thanks to the low power). Using a smartphone to do things is a nowadays trend, devices connected to smartphones have inherited the name "smart" devices even though most of those are functionally terribly stupid. Almost all "smart" devices are BLE connected devices and there is a tsunami of tech startup companies out there making more and more applications of this technology. Smartphone centered wearables and home appliances technology is coming into our lives for good.

IOT: Bridging MCUs to the Internet

BLE has a usable range of about 20 to 50 meters. In the trend of another "smart" system, "smart homes", we usually need to access appliances and equipment from everywhere and accessing from everywhere means Internet. Since computers that can connect to the internet like the Espressif's ESP8266 or the ESP32, they cost little and are super small, they easily embed into devices making the "Internet of Things", an even bigger tech startup tsunami. Almost all use WiFi to connect, since it is now the de facto wireless network to all homes, business and public places. Cellular connection is costly (a SIM with a data plan for each device), 5G is to come at some point in history, prove me wrong, it will fail or delay a lot. Devices connected over the WiFi are surely accessible from the local WiFi network within our home that is also using the internet protocol. To get accessible though out of it they have to connect to a server that is accessible to the internet. Let's see the internet basics.

Internet Basics

It is not easier to write a better introduction than the "Internet Protocol" article from Wikipedia (as of Jan 2020):

The Internet Protocol (IP) is the principal communications protocol in the Internet protocol suite for relaying datagrams across network boundaries. Its routing function enables internetworking, and essentially establishes the Internet.

IP has the task of delivering packets from the source host to the destination host solely based on the IP addresses in the packet headers. For this purpose, IP defines packet structures that encapsulate the data to be delivered. It also defines addressing methods that are used to label the datagram with source and destination information.

Historically, IP was the connectionless datagram service in the original Transmission Control Program introduced by Vint Cerf and Bob Kahn in 1974, which was complemented by a connection-oriented service that became the basis for the Transmission Control Protocol (TCP). The Internet protocol suite is therefore often referred to as TCP/IP.

The first major version of IP, Internet Protocol Version 4 (IPv4), is the dominant protocol of the Internet. Its successor is Internet Protocol Version 6 (IPv6), which has been in increasing deployment on the public Internet since c. 2006.

You can take this on and continue learning, from zero, the whole "inside the cables signals" picture of how internet works. Approaching that knowledge from the MCUs software development, you may use real "just doing the job" web servers, http clients and "sockets" in your MCUs (ESPs are mostly suggested). That way you might make yourself a network engineer and internet apps developer. In this approach you will get to know a lot in internet and network engineering, starting from the ground up, compared to just setting up PCs and servers, not knowing the underneath reality, or compared to making web sites on WordPress or Joomla instead of real HTML and a little of JavaScript. Last, you will make 2-5 internet devices of tiny dimensions interacting with the real world (e.g. measuring and controlling something) instead of interacting with information only.

2.16 FUN HAS JUST BEGUN, WHERE TO GO NEXT

Let us end with some advices from a guy with almost 3 decades of experience.

WHERE WE ARE

Welcome to the world of electronics and software engineering. If you have understood and remember the 100% of this book, having not touched an Arduino with your hands yet, your ranging in the hobbyist world is not bad.

My very subjective view is that among the around 10 million Arduino owners, you rang somewhere near the middle regarding knowledge around the Arduino and above the middle regarding electronics knowledge. If you give a week in making Arduino projects of your choice, on which you may go with self - guidance from now on, your ranking will go up by another 20%. Note here that we are talking in statistics terms, about the 50th percentile, not the 50% of the knowledge!

In regards to where you score among the professional embedded electronics designers you have covered about the 10% of the road in the essential knowledge of hardware and about the 5% in knowledge of software. In experience, regarding PCBs design (design tools using also) and building code of big projects (not what the command x does but building big projects using it), you are at 0% and that road is really long.

Good news is that you will understand the 90% of the terminology, so your communication with professionals will be awesome. As a hobbyist all the knowledge you find in the internet will be in a comprehensive language. In case you are a manager who previously didn't know of electronics and software but you are engaged in electronics projects, after this book the guys in the team will stop sounding "Greek" in your ears anymore. On top of that you will actually feel like one of them.

ROADS UNWINDING AHEAD

You may go from hobbyist to product designer and undertake really challenging and actual design missions with beers paid. That road is to be taken with little steps knowing there is a long, long way to go starting with some fails probably and moderate outcome some other times. So you have to commit to customers contracts when you count already quite a few projects made by you for you, having completed at least a few that count over 1000 lines of code. Regarding hardware design, you should count a dozen of PCBs designed by you and made by you at least.

Should you graduate a school/college/university for that? To my opinion you don't have to. Many in this field (all are among the top ones) have no such specialty degree, especially the software developers. By the way, if you dream to follow a career as a software developer it is not a bad method to begin with Arduinos. Software here works in a machine (MCU) that you can easily get to know at its 100%, unlike the big personal computers with thousands of CPUs in the graphics card and terribly complex structure and features in their hardware and in their operating system.

BEYOND ARDUINO

This book was not starting and ending in a few Arduino projects. It started electronics from the ground up, giving the background in electronic theory, components, techniques and finally in MCUs technology. The reason was to show to you a horizon that goes further than Arduinos. After some Arduino projects you may fly to worlds of non-Arduino MCUs, if of course you ever feel like it and there is a benefit there to you. A good starting point of non-Arduino worlds may be STM32 MCUs. Prepare to program in C at most of those MCU universes, and to take time to learn the full C language and to understand and use more complex libraries (APIs) than the

Arduino ecosystem. The benefits: a) Walk on the long road that goes from hobbyist to product designer. b) In non-Arduino worlds you will usually find in-circuit debugging, seamless watching of variable's values on run-time, multi-tasking operating systems, MCUs that fit exactly to your project's needs and many more goodies. The cons: Less "ready to use" code, you may have to write more, find "ready to use" less.

BEYOND ARDUINO IDE

The IDE we have been using is designed only around the concept "we must not scare newcomers". It is very primitive in general regarding facilities other nowadays IDEs provide. You should look about platformIO in Micorsoft's "Visual Studio Code" in a video of Andreas Spiess in YouTube (searching there for with those keywords) after a few projects in the Arduino's IDE. You will be really amazed on how better you can program in such IDEs with automatic code completion and error highlighting as you type. Even while you are still at the Arduino IDE you surely must have and use the notepad++ code editor (free), at least for "opening" other's people code on your PC.

RECOMMENDED READING

- Reading in the internet about whatever matter comes to your path. Try to separate beginner's conversations on blogs from serious great articles on web sites like arduino.cc sparkfun.com, adafruit.com, manufacturer's articles in sort pdfs and other super great sources.
- Datasheets and much searching in digikey.com and LCSC.com for components.
- A good C language book (better leave C++ for a next step) like the C's creator Dennis Ritchie "C Programming Language, 2nd Edition" that is the full language reference and really well written.

- The Art of Electronics (3rd edition). The bible of electronics, really. Like in this book, almost no math, practical to the bone. It is the most influential book to writing the one you are reading now. Not a lot in MCUs knowledge, but it covers electronics engineering knowledge from ground up to the sky.

You need to combine two ways of learning. Reading books for acquiring general and structured knowledge and learning over googling for a specific problem you fall into, or "learn as you make". The last is far more efficient but the first is needed also. Have the "learn as you make" as the first choice and apply it for about the 70% of the time devoted to learning.

The information out there is huge. Try to ask the best questions first and then get answers at a skin level in many cases. Do not approach learning as a road where you leave no gaps behind. Try to fly over and see the big picture the most that you can. Read deeply about matters that you are busy with at the moment. Know that experience is more valuable at many cases (like PCBs design and software development) so make more and read less.

RECOMMENDED YOUTUBE MATERIAL

In my view, videos are to get experience about a subject mostly. If a "how to" or "what is" is what you are missing, reading will get you there 2 or 5 times faster. But videos are a treasure in getting experience of hardware without touching it, or seeing how well a software library is working. And yes, you have a great relaxing time too.

Here are some greatly recommended channels (as of January 2020):

- ElectroBOOM: You get to love electrical engineering by this guy!! A great electronics influencer to persons not involved to electronics, beginners and even experts.
- GreatScott! Speechless of his great clarity and in-depth analysis
- Andreas Spiess: The guy with the Swiss accent who rocks!

- educ8s.tv: The guy with the Greek accent who presents every topic awesomely.
- Afrotechmods: for great intro videos.
- EEVblog: for real and I mean real experience and expertise on over 1000 topics. Most are in very advanced level.

But I am sure I have missed out a lot. Sorry if I do not have some great guys in that very sort list.

Making vs Passively Learning

Learning how to drive by books and driving is the difference of learning vs doing here, especially in software development and in PCBs design (if you start on that).

A recommendation: Type yourself the first programs of the chapter 2.14 and play with them on your Arduino. Purposely they are not on the internet for copy-pasting them. You should train in finding typo errors, use tabs properly, and have the feeling of incrementally making, not copying.

The Spectrum of Engineering Skills

In programmers who are employed in big companies, the best ones are more than 10 times more productive than the least capable guys in the same company. In electronics design the situation is the same and worse. The great ones I have met all and I mean all, love what they do and do it as a hobby besides their employment time. You get good to what attracts you and joys you.

Conclusions

Hope you enjoyed. Opening a door in your life had been my goal. Walking through is up to you if there is meaning there for you.

Enjoy making your ideas.

YOU CAN REACH ME:

- Making a highly appreciated Amazon review. Next edition (2021) will get better mostly by listening to your precious like gold feedback, so you could make next readers happier.
- Feel free to discuss any topic with me or others on the blog / forum for this book. Go to the site www.jfragos.com or "Google" the title of the book + the word "forum" and you will get there easily enough.
- If you happen to read this from a digital copy circulating on a torrent (not having payed moneys for it) and you feel like buying me a small beer for my work, it is just a very few clicks on Paypal at *jimfragos@gmail.com* or just buy it from Amazon, and I will shout a loud "Cheers"

3. APPENDIX

3.1 MULTIPLIER PREFIXES

Where may zeros are involved, e.g. 0.000000021 or 13200000 instead of having to count zeros each time on such numbers, in engineering terminology we use some multipliers named in easy names to remember. Here they are:

Acronym	Symbol	Value
tera	T	x 1 000 000 000 0000
giga	G	x 1 000 000 000
mega	M	x 1 000 000
kilo	K	x 1 000
milli	m	x 0.001
micro	µ or u	x 0.000 001
nano	n	x 0.000 000 001
pico	p	x 0.000 000 000 001

Examples:

- A resistor of 1200 Ohms is 1.2Kohms or in sort 1.2K or 1K2
- The WiFi radio frequency is around 2400000Hz that is 2.4GHz
- A capacitor 0.0000001 Farad is 100nF or 0.1uF

3.2 A REAL DATASHEET OF AN LED

**DOUBLE HETEROJUNCTION AlGaAs
LOW CURRENT RED LED LAMPS**

T-1 3/4 (5mm)	HLMP-D150A	Red Diffused
	HLMP-D155A	Red Clear with Standoff
T-100 (3mm)	HLMP-K150	Red Diffused
	HLMP-K155	Red Clear

PACKAGE DIMENSIONS

FEATURES
- Wide Viewing Angle
- Deep Red Color

DESCRIPTION

Exceptional light output typifies these devices and provides for their use over a broad range of drive currents. The LED material is based on double heterojunction (DH) AlGaAs/GaAs technology.

NOTES:

1. ALL DIMENSIONS ARE IN INCHES (mm).
2. TOLERANCE ARE ±.010" UNLESS OTHERWISE SPECIFIED.
3. AN EPOXY MENISCUS MAY EXTEND ABOUT .040"(1 mm) DOWN THE LEADS.

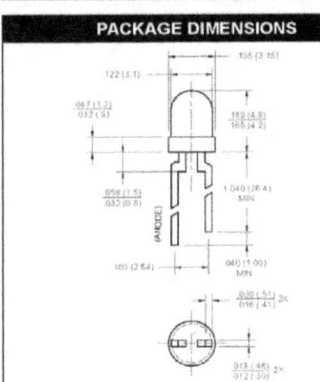

HLMP-K150/K155

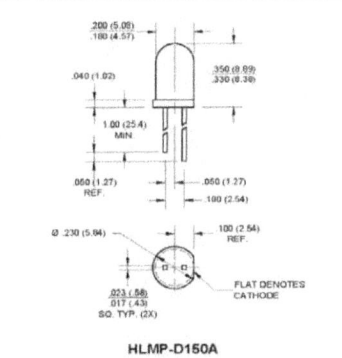

HLMP-D150A

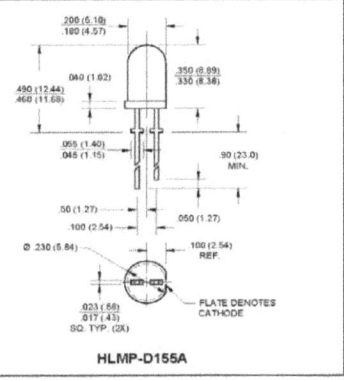

HLMP-D155A

DOUBLE HETEROJUNCTION AIGaAs
LOW CURRENT RED LED LAMPS

ABSOLUTE MAXIMUM RATING (T$_A$ =25°C)

Parameter	RED	UNITS
Power Dissipation	87	mW
Peak Forward Current (f=1kHz, DF=10%)	300	mA
Continuous DC Forward Current	30	mA
Lead Soldering Time at 260° C	5	sec
Operating Temperature	-20 to +100	°C
Storage Temperature	-55 to +100	°C

ELECTRICAL / OPTICAL CHARACTERISTICS (T$_A$ =25°C)

Parameter	HLMP-K150	HLMP-K155	HLMP-D150A	HLMP-D155A	Condition
Luminous Intensity (mcd)					I_F = 1mA
Minimum	1.2	2.0	1.2	3.0	
Typical	2.0	3.0	3.0	10.0	
Forward Voltage (V)					I_F = 1mA
Maximum	1.8	1.8	1.8	1.8	
Typical	1.6	1.6	1.6	1.6	
Peak Wavelength (nm)	660	660	660	660	I_F = 1mA
Spectral Line Half Width	20	20	20	20	I_F = 1mA
Reverse Voltage (V)	5	5	5	5	I_R = 100µA
Viewing Angle (°)	60	45	65	24	I_F = 1mA

 **DOUBLE HETEROJUNCTION AlGaAs LOW CURRENT RED LED LAMPS**

TYPICAL PERFORMANCE CURVES ($T_A = 25°C$)

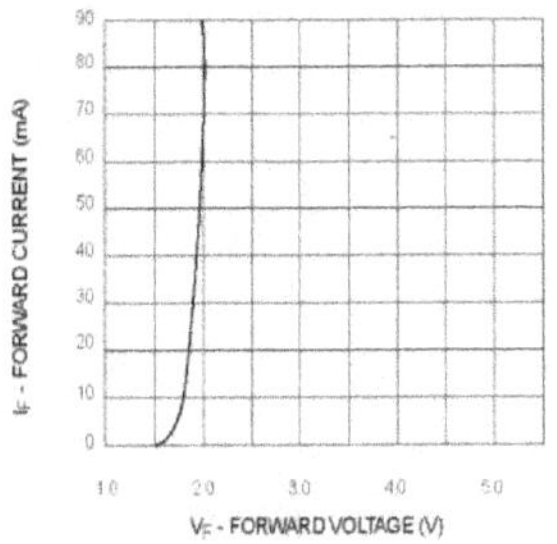

Fig. 1 Forward Current vs. Forward Voltage

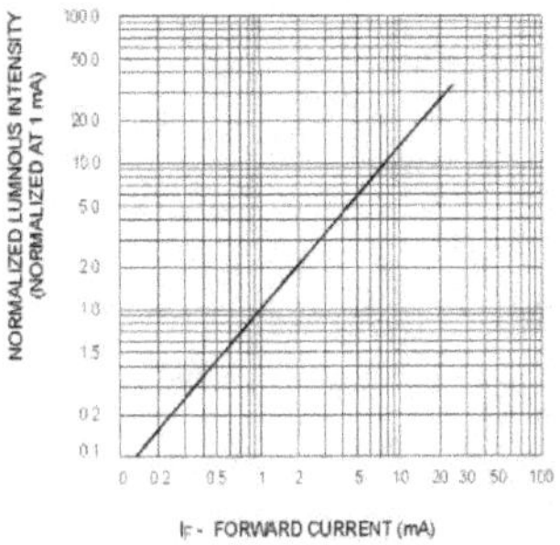

Fig. 2 Relative Luminous Intensity vs. DC Forward Current

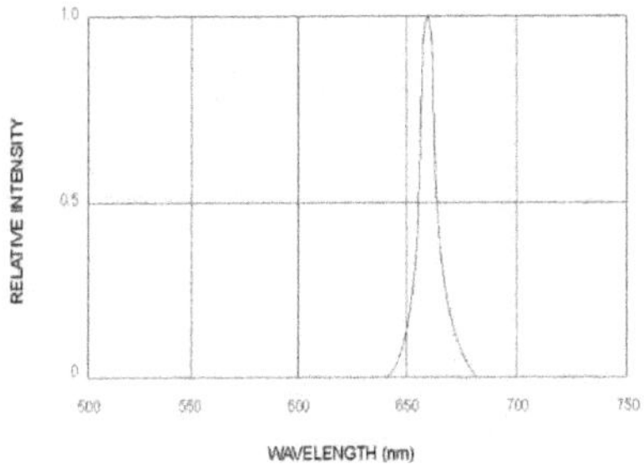

Fig. 3 Relative Intensity vs. Peak Wavelength

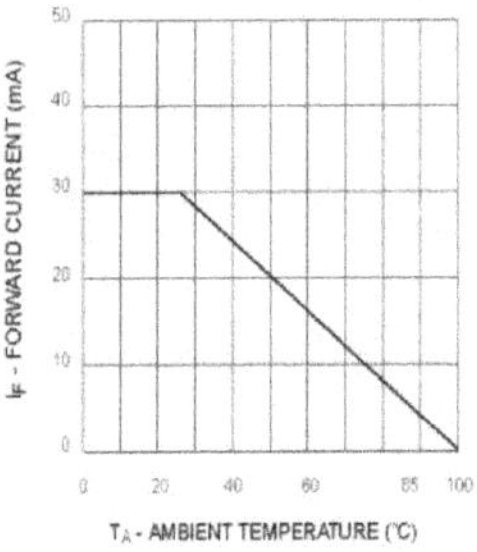

Fig. 4 Current Derating Curve

EVERLIGHT

DOUBLE HETEROJUNCTION AIGaAs LOW CURRENT RED LED LAMPS

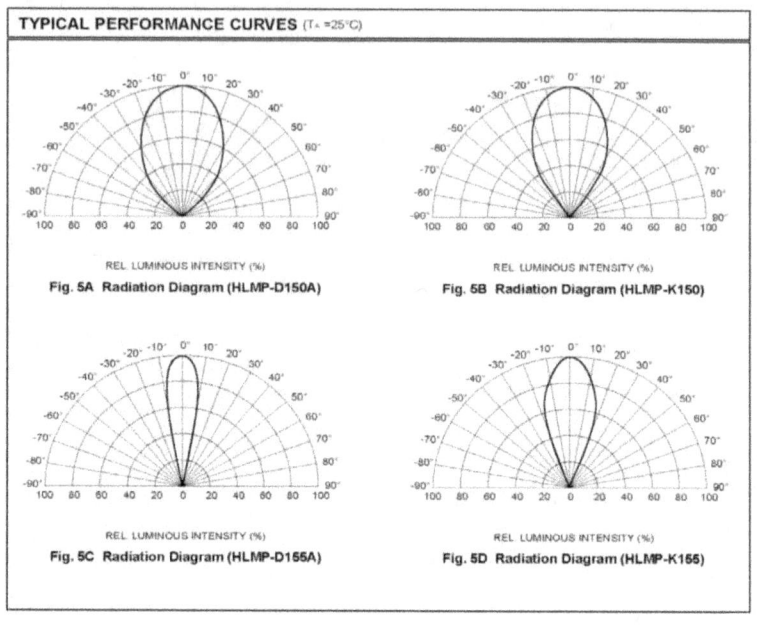

TYPICAL PERFORMANCE CURVES (T_A =25°C)

Fig. 5A Radiation Diagram (HLMP-D150A)

Fig. 5B Radiation Diagram (HLMP-K150)

Fig. 5C Radiation Diagram (HLMP-D155A)

Fig. 5D Radiation Diagram (HLMP-K155)

3.3 DATASHEET HIGHLIGHTS OF SOME NOTABLE MCUS

STM32F030 SERIES

**STM32F030x4 STM32F030x6
STM32F030x8 STM32F030xC**

Value-line Arm®-based 32-bit MCU with up to 256 KB Flash, timers, ADC, communication interfaces, 2.4-3.6 V operation

Datasheet - production data

Features

- Core: Arm® 32-bit Cortex®-M0 CPU, frequency up to 48 MHz
- Memories
  - 16 to 256 Kbytes of Flash memory
  - 4 to 32 Kbytes of SRAM with HW parity
- CRC calculation unit
- Reset and power management
  - Digital & I/Os supply: V_{DD} = 2.4 V to 3.6 V
  - Analog supply: V_{DDA} = V_{DD} to 3.6 V
  - Power-on/Power down reset (POR/PDR)
  - Low power modes: Sleep, Stop, Standby
- Clock management
  - 4 to 32 MHz crystal oscillator
  - 32 kHz oscillator for RTC with calibration
  - Internal 8 MHz RC with x6 PLL option
  - Internal 40 kHz RC oscillator
- Up to 55 fast I/Os
  - All mappable on external interrupt vectors
  - Up to 55 I/Os with 5V tolerant capability
- 5-channel DMA controller
- One 12-bit, 1.0 µs ADC (up to 16 channels)
  - Conversion range: 0 to 3.6 V
  - Separate analog supply: 2.4 V to 3.6 V
- Calendar RTC with alarm and periodic wakeup from Stop/Standby
- 11 timers
  - One 16-bit advanced-control timer for six-channel PWM output
  - Up to seven 16-bit timers, with up to four IC/OC, OCN, usable for IR control decoding
  - Independent and system watchdog timers
  - SysTick timer

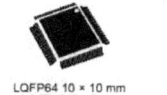
LQFP64 10 × 10 mm
LQFP48 7 × 7 mm
LQFP32 7 × 7 mm

TSSOP20 6.4 × 4.4 mm

- Communication interfaces
  - Up to two I²C interfaces
    - Fast Mode Plus (1 Mbit/s) support on one or two I/Fs, with 20 mA current sink
    - SMBus/PMBus support (on single I/F)
  - Up to six USARTs supporting master synchronous SPI and modem control; one with auto baud rate detection
  - Up to two SPIs (18 Mbit/s) with 4 to 16 programmable bit frames
- Serial wire debug (SWD)
- All packages ECOPACK®2

Table 1. Device summary

Reference	Part number
STM32F030x4	STM32F030F4
STM32F030x6	STM32F030C6, STM32F030K6
STM32F030x8	STM32F030C8, STM32F030R8
STM32F030xC	STM32F030CC, STM32F030RC

Description

STM32F030x4/x6/x8/xC

Table 2. STM32F030x4/x6/x8/xC family device features and peripheral counts

Peripheral		STM32 F030F4	STM32 F030K6	STM32 F030C6	STM32 F030C8	STM32 F030CC	STM32 F030R8	STM32 F030RC
Flash (Kbytes)		16	32	32	64	256	64	256
SRAM (Kbytes)		4	4	4	8	32	8	32
Timers	Advanced control	1 (16-bit)						
	General purpose	4 (16-bit)[1]			5 (16-bit)			
	Basic	-	-	-	1 (16-bit)[2]	2 (16-bit)	1 (16-bit)[2]	2 (16-bit)
Comm. interfaces	SPI	1[3]	1[3]	1[3]	2	2	2	2
	I²C	1[4]	1[4]	1[4]	2	2	2	2
	USART	1[5]	1[5]	1[5]	2[6]	6	2[6]	6
12-bit ADC (number of channels)		1 (9 ext. +2 int.)	1 (10 ext. +2 int.)	1 (10 ext. +2 int.)	1 (10 ext. +2 int.)	1 (10 ext. +2 int.)	1 (16 ext. +2 int.)	1 (16 ext. +2 int.)
GPIOs		15	26	39	39	37	55	51
Max. CPU frequency		48 MHz						
Operating voltage		2.4 to 3.6 V						
Operating temperature		Ambient operating temperature: -40°C to 85°C / Junction temperature: -40°C to 105°C						
Packages		TSSOP20	LQFP32	LQFP48			LQFP64	

1. TIM15 is not present.
2. TIM7 is not present.
3. SPI2 is not present.
4. I2C2 is not present.
5. USART2 to USART6 are not present.
6. USART3 to USART6 are not present

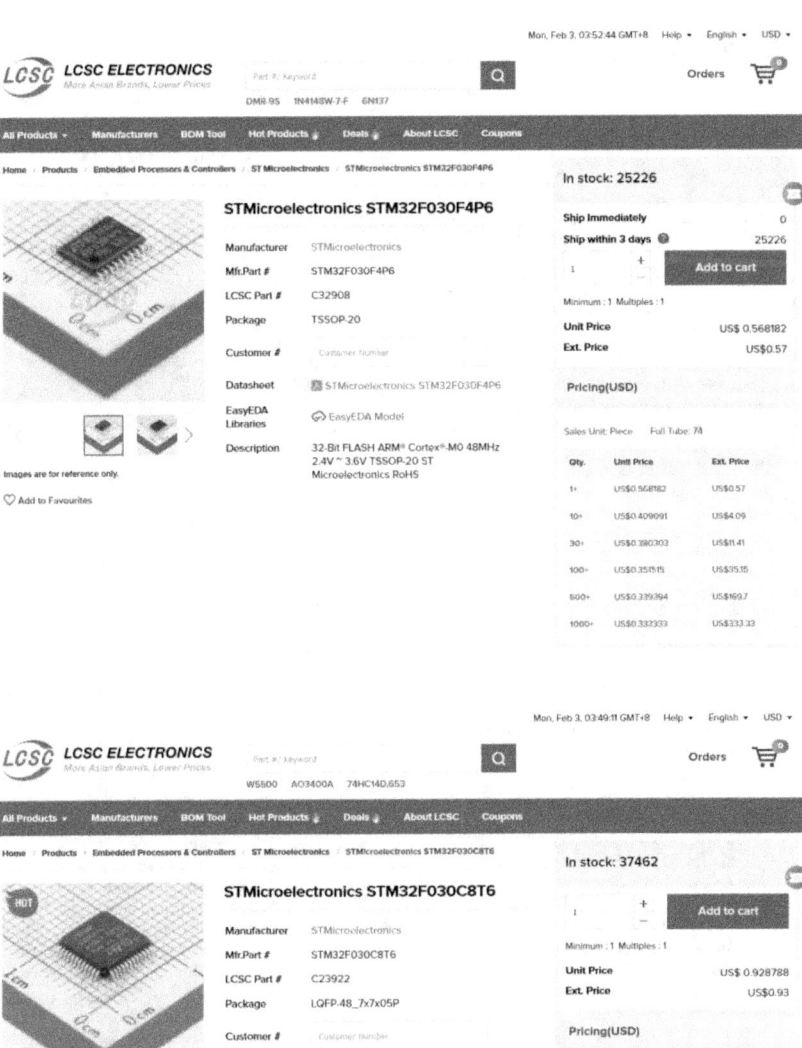

STM32F4 SERIES

STM32F446xC/E

ARM® Cortex®-M4 32b MCU+FPU, 225DMIPS, up to 512kB Flash/128+4KB RAM, USB OTG HS/FS, 17 TIMs, 3 ADCs, 20 comm. interfaces

Datasheet - production data

Features

LQFP64 (10 × 10mm)
LQFP100 (14 × 14mm) UFBGA144 (7 x 7 mm)
LQFP144 (20 x 20 mm) UFBGA144 (10 x 10 mm) WLCSP 81

- Core: ARM® 32-bit Cortex®-M4 CPU with FPU, Adaptive real-time accelerator (ART Accelerator™) allowing 0-wait state execution from Fl ash memory, frequency up to 180 MHz, MPU, 225 DMIPS/1.25 DMIPS/MHz (Dhrystone 2.1), and DSP instructions
- Memories
  - 512 kB of Flash memory
  - 128 KB of SRAM
  - Flexible external memory controller with up to 16-bit data bus: SRAM,PSRAM,SDRAM/LPSDR SDRAM, Flash NOR/NAND memories
  - Dual mode Quad SPI interface
- LCD parallel interface, 8080/6800 modes
- Clock, reset and supply management
  - 1.7 V to 3.6 V application supply and I/Os
  - POR, PDR, PVD and BOR
  - 4-to-26 MHz crystal oscillator
  - Internal 16 MHz factory-trimmed RC (1% accuracy)
  - 32 kHz oscillator for RTC with calibration
  - Internal 32 kHz RC with calibration
- Low power
  - Sleep, Stop and Standby modes
  - V_{BAT} supply for RTC, 20×32 bit backup registers + optional 4 KB backup SRAM
- 3×12-bit, 2.4 MSPS ADC: up to 24 channels and 7.2 MSPS in triple interleaved mode
- 2×12-bit D/A converters
- General-purpose DMA: 16-stream DMA controller with FIFOs and burst support
- Up to 17 timers: 2x watchdog, 1x SysTick timer and up to twelve 16-bit and two 32-bit timers up to 180 MHz, each with up to 4 IC/OC/PWM or pulse counter
- Debug mode
  - SWD & JTAG interfaces
  - Cortex®-M4 Trace Macrocell™

- Up to 114 I/O ports with interrupt capability
  - Up to 111 fast I/Os up to 90 MHz
  - Up to 112 5 V-tolerant I/Os
- Up to 20 communication interfaces
  - SPDIF-Rx
  - Up to 4 × I^2C interfaces (SMBus/PMBus)
  - Up to 4 USARTs/2 UARTs (11.25 Mbit/s, ISO7816 interface, LIN, IrDA, modem control)
  - Up to 4 SPIs (45 Mbits/s), 3 with muxed I^2S for audio class accuracy via internal audio PLL or external clock
  - 2 x SAI (serial audio interface)
  - 2 × CAN (2.0B Active)
  - SDIO interface
  - Consumer electronics control (CEC) I/F
- Advanced connectivity
  - USB 2.0 full-speed device/host/OTG controller with on-chip PHY
  - USB 2.0 high-speed/full-speed device/host/OTG controller with dedicated DMA, on-chip full-speed PHY and ULPI
  - Dedicated USB power rail enabling on-chip PHYs operation throughout the entire MCU power supply range
- 8- to 14-bit parallel camera interface up to 54 Mbytes/s
- CRC calculation unit
- RTC: subsecond accuracy, hardware calendar
- 96-bit unique ID

Table 1. Device summary

Reference	Part number
STM32F446xC/E	STM32F446MC, STM32F446ME, STM32F446RC, STM32F446RE, STM32F446VC, STM32F446VE, STM32F446ZC, STM32F446ZE

September 2016   DocID027107 Rev 6

3.3 Datasheet highlights of some notable MCUs

ESP32

1.4 MCU and Advanced Features

1.4.1 CPU and Memory

- Xtensa® single-/dual-core 32-bit LX6 microprocessor(s), up to 600 MIPS (200 MIPS for ESP32-S0WD, 400 MIPS for ESP32-D2WD)
- 448 KB ROM
- 520 KB SRAM
- 16 KB SRAM in RTC
- QSPI supports multiple flash/SRAM chips

1.4.2 Clocks and Timers

- Internal 8 MHz oscillator with calibration
- Internal RC oscillator with calibration
- External 2 MHz ~ 60 MHz crystal oscillator (40 MHz only for Wi-Fi/BT functionality)
- External 32 kHz crystal oscillator for RTC with calibration
- Two timer groups, including 2 × 64-bit timers and 1 × main watchdog in each group
- One RTC timer
- RTC watchdog

1.4.3 Advanced Peripheral Interfaces

- 34 × programmable GPIOs
- 12-bit SAR ADC up to 18 channels
- 2 × 8-bit DAC
- 10 × touch sensors
- 4 × SPI
- 2 × I²S
- 2 × I²C
- 3 × UART
- 1 host (SD/eMMC/SDIO)
- 1 slave (SDIO/SPI)
- Ethernet MAC interface with dedicated DMA and IEEE 1588 support
- CAN 2.0
- IR (TX/RX)
- Motor PWM
- LED PWM up to 16 channels
- Hall sensor

1.4.4 Security

- Secure boot
- Flash encryption
- 1024-bit OTP, up to 768-bit for customers
- Cryptographic hardware acceleration:
    - AES
    - Hash (SHA-2)
    - RSA
    - ECC
    - Random Number Generator (RNG)

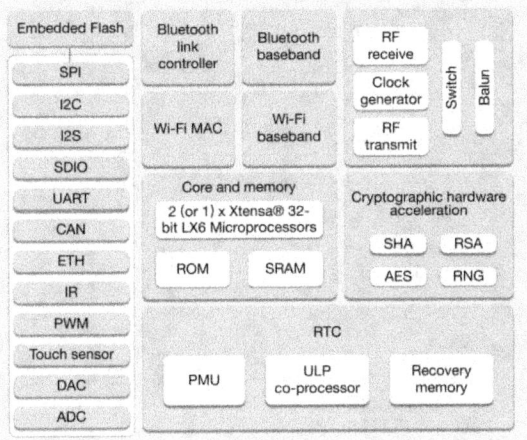

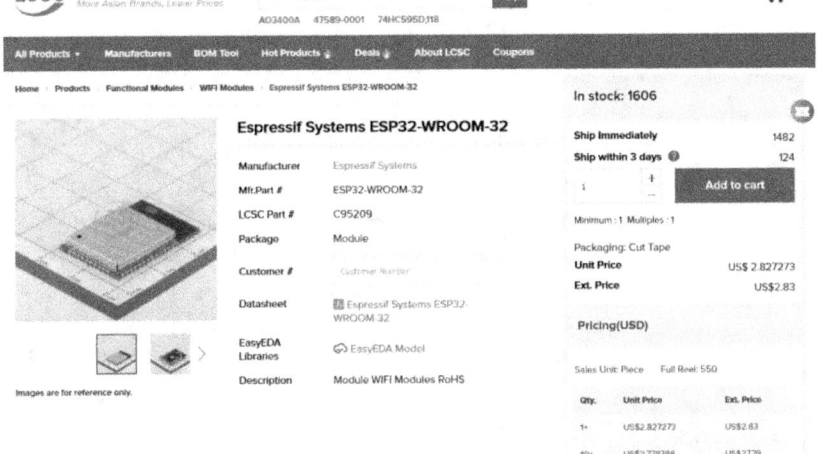

www.ingramcontent.com/pod-product-compliance
Lightning Source LLC
Chambersburg PA
CBHW070620220526
45466CB00001B/67